Advances and Applications in Biotechnology

Advances and Applications in Biotechnology: Volume 1—Prospects of Farm Animals and Future Developments explores the dynamic and evolving role of livestock in global agriculture, emphasizing sustainable practices, molecular innovations, and emerging technologies. Against the backdrop of increasing environmental pressures and rising food security concerns, this volume highlights the transformative potential of biotechnology in farm animal research and management.

This book aims to integrate traditional livestock knowledge with advanced scientific strategies, offering readers a multidimensional view of animal agriculture. It addresses challenges in genetics, health, nutrition, welfare, climate resilience, and productivity, fostering dialogue between researchers, technologists, and practitioners.

Encompassing a broad scope—from glycoproteomics and immunity to species-specific developments—this volume navigates cutting-edge discoveries and practical applications across farm animal systems. Contributors offer global perspectives rooted in rigorous research and collaborative expertise.

Key Features

- Showcases molecular and biotechnological tools in farm animal improvement.
- Discusses immune resilience, genomic adaptations, and nutritional strategies.
- Covers environmental impacts and sustainable livestock frameworks.
- Features species-specific chapters with empirical and applied insights.
- Highlights case studies and data-driven approaches for problem-solving.

This book is a valuable resource for researchers, veterinarians, students, agritech professionals, and policymakers. By bridging foundational biology with futuristic approaches, it encourages innovation, inclusivity, and resilience in livestock science and global agricultural systems.

Prospect of Animal Biotechnology and Future Developments
Series Editor: Arti Gupta

Advances and Applications in Biotechnology: Volume 1—Prospects of Farm Animals and Future Developments
Edited by Arti Gupta and Ali Asghar Rastegari

Advances and Applications in Biotechnology

Volume 1 — Prospects of Farm Animals and Future Developments

Edited by

Arti Gupta and Ali Asghar Rastegari

CRC Press
Taylor & Francis Group
Boca Raton London New York

CRC Press is an imprint of the
Taylor & Francis Group, an **informa** business

First edition published 2026
by CRC Press
2385 NW Executive Center Drive, Suite 320, Boca Raton FL 33431

and by CRC Press
4 Park Square, Milton Park, Abingdon, Oxon, OX14 4RN

CRC Press is an imprint of Taylor & Francis Group, LLC

ISBN: 9781032590745 (hbk)
ISBN: 9781041205470 (pbk)
ISBN: 9781003452805 (ebk)

DOI: 10.1201/9781003452805

Typeset in Times
by codeMantra

Contents

Preface

This volume offers a timely exploration of the evolving interface between biotechnology and animal husbandry. As the global demand for sustainable agricultural practices intensifies, this volume delves into innovative biotechnological approaches that promise to reshape how farm animals are bred, nurtured, and integrated into broader agri-food systems.

Bringing together contributions from scholars across molecular biology, microbiology, genetics, and environmental science, this volume presents an in-depth overview of technological interventions that aim to improve livestock productivity, animal welfare, and ecological balance. Key topics include breeding advancements, artificial insemination (AI) techniques, and the strategic role of biotechnology in addressing climate-resilient food systems. This work fosters further research aimed chapter-wise at harmonizing technology and sustainability in animal agriculture.

Chapter 1 explores the genetic foundations of farm animal improvement and how biotechnological tools are revolutionizing breeding strategies for enhanced traits and disease resistance.

Chapter 2 reviews the development of AI technologies, with a focus on species-specific adaptations, and highlights the increasing precision of reproductive biotechnology.

Chapter 3 provides a futuristic lens on AI applications, including advanced semen preservation, synchronization techniques, and their broader implications for animal productivity.

Chapter 4 analyzes biotechnology's critical role in aligning livestock management with global sustainability goals, including climate resilience and food security.

Chapter 5 offers a comparative review of innovations across plant and animal biotechnology, drawing connections between crop enhancement and livestock integration.

Chapter 6 outlines strategic pathways and policy implications for harnessing biotechnology, addressing regulatory hurdles and ethical considerations.

Chapter 7 highlights the ecological and economic merits of integrated systems, emphasizing biotechnology's role in optimizing cross-species resource efficiency.

Chapter 8 investigates quality assurance protocols vital for sustainable biotechnological practices, especially under evolving environmental compliance frameworks.

Chapter 9 focuses on novel feed solutions derived through biotechnology, with attention to nutrient efficiency, safety standards, and waste reduction.

Chapter 10 concludes with a philosophical and scientific inquiry into the ethical and existential dimensions of farming animals in a biotechnological era.

Together, these chapters present a forward-looking narrative for researchers, practitioners, and policymakers invested in building a sustainable and innovative future for agriculture and livestock science.

Acknowledgments

First and foremost, we would like to thank God for his never-ending grace, mercy, and provision during what ended up being one of the toughest times of our lives.

We are indebted to the many people who helped bring this book to light. We wish to thank Gagandeep Singh, Senior Publisher; Dr. Renu Upadhyay, Commissioning Editor; Tisha Sharma, Editorial Assistant; and Taylor & Francis/CRC Press for their generous assistance, constant support, and patience in initiating the series volume. The editors are very thankful to the Vice Chancellor, Dr. Ram Manohar Lohia Avadh University, Ayodhya, India, and Islamic Azad University, Iran, for their kind support and constant encouragement. Special thanks are due to well-wishers, family members, and friends.

Dr. Arti Gupta owes a debt of gratitude to her wonderful husband, Mr. Saket Agrawal, and their son, Reyansh. Their unrelenting support and motivation have been pivotal in inspiring her to undertake this endeavour, overcome all obstacles, and assemble all the pieces into a coherent whole. Dr. Ali gives his special regards to his family for their constant, invaluable support and motivation to publish this book. We also give special thanks to all faculty colleagues and senior faculty members of our team.

Arti Gupta
Dr. Ram Manohar Lohia Avadh University, India

Ali Asghar Rastegari
Islamic Azad University, Iran

About the Editors

Arti Gupta earned her Ph.D. in Animal Science from Mahatma Jyotiba Phule Rohilkhand University in Bareilly, India, in 2010. Her current research interests include animal biotechnology, molecular plant biotechnology, molecular animal biotechnology, bioprocess technology, and microbiology. Since July 2014, she has been serving at Shri Avadh Raj Singh Smarak Degree College in Gonda, Uttar Pradesh, India, where she is also engaged in editorial work.

With 13 years of teaching and research experience in plant and animal biotechnology and microbial biotechnology, she has received several awards, including the University Topper (Gold Medal) for her M.Sc. in Biotechnology, the Young Scientist Award (Gold Medal), Best Poster Presenter, and the Dr. V.P. Agarwal Award (Gold Medal).

She has contributed extensively to academia, with two monographs, 21 edited books published by various international publishers, 16 book chapters, and approximately 55 national and international research papers. She holds lifetime memberships in the Indian Science Congress Association, the Biotech Research Society of India, the Zoological Society of India, and the International Consortium of Contemporary Biologists. Additionally, she serves on several editorial boards.

Previously, she was a member of the Teaching Personnel at G.B. Pantnagar University, Pantnagar, Uttarakhand, India, and served as a visiting academic researcher at the Biorefining and Advanced Material Research Center (BAMRC) at Scotland's Rural College (SRUC), University of Edinburgh, United Kingdom.

Ali Asghar Rastegari was born in 1962 and is currently a retired Associate Professor in the Department of Biochemistry, Islamic Azad University, Isfahan, Iran. He has over many years of research and teaching experience in the field of enzyme biotechnology/nanobiotechnology, biophysical chemistry, computational biology, and biomedicine. He has published nine books and 13 book chapters with various reputed international publishers. He is a reviewer for different international journals.

Contributors

Mohamed Morsi Mohamed Ahmed
Nucleic Acids Research Department
Genetic Engineering and Biotechnology Research Institute
Alexandria, Egypt

Amro Abd Al Fattah Amara
Protein Research Department
Genetic Engineering and Biotechnology Research Institute
New Borg El-Arab, Alexandria, Egypt

Ranjitha Ashok
Department of Biotechnology
Maharani Lakshmi Ammanni College for Women Autonomous
Malleshwaram, Bengaluru, Karnataka, India

Nawal Abd El-Baky
Protein Research Department
Genetic Engineering and Biotechnology Research Institute
New Borg El-Arab, Egypt

Mounir Mohamed Salem-Bekhit
Department of Pharmaceutics, College of Pharmacy
King Saud University
Riyadh, Saudi Arabia

Jolitha Adyar Bose
Department of Biotechnology
Maharani Lakshmi Ammanni College for Women Autonomous
Malleshwaram, Bengaluru, Karnataka

Lipsa Dehal
Department of Zoology & Environmental Sciences
Punjabi University
Patiala, India

Rekha Dhanai
Department of Agriculture
Tula's Institute
Dehradun, India

Renita Dsouza
Department of Biotechnology
St Aloysius (Deemed to be University)
Mangaluru, India

D. Geetha
Department of Biotechnology
Maharani Lakshmi Ammanni College for Women Autonomous
Malleshwaram, Bengaluru, Karnataka, India

Arti Gupta
Department of Zoology, Shri Avadh Raj Singh Smarak Degree College
Dr. Ram Manohar Lohia Avadh University
Ayodhya, India

Tank Sinderpal Kabalsingh
Department of Chemistry
Kanya Maha Vidyalaya
Jalandhar, India

Mandeep Kaur
Department of Zoology
Kanya Maha Vidyalaya
Jalandhar, India

Shweta Maheshwari
Department of Zoology
Kanya Maha Vidyalaya
Jalandhar, India

Jyothi Miranda
Department of Botany
St Aloysius (Deemed to be University)
Mangaluru, India

Kapil Paul
Department of Zoology
Kanya Maha Vidyalaya
Jalandhar, India

Pahul Padam
Department of Soft Skills, Centre for Professional Enhancement
Lovely Professional University
Phagwara, India

Prabha
Department of Zoology
Kanya Maha Vidyalaya
Jalandhar, India

Surbhi Pradhan
Department of Medical Laboratory Technology and Medical Microbiology
Sardar Bhagwan Singh University
Dehradun, India

Palani Preetham
Department of Biotechnology
Maharani Lakshmi Ammanni College for Women Autonomous
Malleshwaram, Bengaluru, Karnataka, India

Shameena Karnad Abdul Rahiman
Department of Biochemistry
St Aloysius (Deemed to be University)
Mangaluru, India

Archana Saini
Department of Zoology
Kanya Maha Vidyalaya
Jalandhar, India

Sana Sheikh
Department of Botany
St Aloysius (Deemed to be University)
Mangaluru, India

Sareen Sheikh
Department of Chemistry,
Yenepoya Institute of Arts, Science, Commerce and Management
Mangaluru, India

Hariprasad Shetty
Department of Zoology
St Aloysius (Deemed to be University)
Mangaluru, India

Khairy Mohamed Abdallah Zoheir
Cell Biology Department
Biotechnology Research Institute, National Research Center
Dokki, Cairo, Egypt

1 Farm Animal Biotechnology as a Driver of Change

Breeding and Genetics

Archana Saini, Lipsa Dehal, and Pahul Padam

1.1 INTRODUCTION

About 12,000–15,000 years ago in the Middle East, people began livestock domestication, which was important for human development and is the keystone to current farming practices. At the start, people chose animals for domestication based on their appearance or traits. Those traits had to suit a local climate and a production setup. Later, in the 16th century, statistical methods were created, which allowed a numeric method for selective breeding for selected traits. This led to the innovation of breeding systems founded on science. These systems are still in operation today. With dependable DNA analysis available, this numeric method has beenextendedto DNA-based breeding concepts. They allow a more cost-effective but still numeric calculation of an animal's genomic breeding value (Todd 2013).

The latter part of the 20th century brought major progress in biological sciences. This progress made embryo transfer (ET) possible. This method allowed the use of the female genetic pool for the first time (Miller 2022). Better cell and embryo culture, gonadotropins that are pure, and the knowledge of the reproductive cycle and hormonal control all affected this technique greatly. ET now sees more use in advanced breeding programs, especially in the top 1%–2% of a given cattle population. Endeavors toward better breeding technologies will not come to fruition until ET is fully implemented and its true impact will not be felt until then (Gullo et al. 2023).

Sexing, cloning, transgenesis, and in vitro embryogenesis all progressed because of ET. A century-old biological idea, that a differentiated cell cannot change into a pluripotent state, was proved wrong with the birth of "Dolly," a cloned sheep, in 1996 (Gardon and Satoe 2020). Recent progress in molecular cell biology has made existing methods effective enough for somatic cloning in the main farm animal species for commercial purposes. This allows the genetic modification (GM) of farm animals with precise changes at a high level of efficiency; it enables the production of new germlines for agricultural as well as biomedical uses. This will speed up the genetic gain in herds or flocks. Precision breeding concepts can now actively be introduced in animal breeding (Khalil 2025). The necessity to strike a balance between the growing demands brought on by the world's population expansion and

DOI: 10.1201/9781003452805-1

the limitations of ecological boundaries and scarce agricultural land makes animal breeding extremely difficult (Bai and Plastow 2022).

Pigs, cattle, sheep, goats, and chickens are examples of farm animals that offer animal-based protein and fat required in daily diet of people worldwide. The demand for livestock products is rising dramatically on a global scale, driven by multiple factors of population increase, urbanization, and rising affluence. A billion people worldwide still suffer from chronic malnutrition, despite significant advancements in agricultural and animal husbandry (Godfray et al. 2010). The lack of livestock protein production will be made worse by global climate change, and existing efforts to meet the world's food demand have also made the already stressed environment worse (Foley et al. 2011). By the middle of this century, the United Nations projects that there will be 10 billion people on the planet. The purpose of breeding has evolved into producing farm animals that have quick growth, excellent disease resistance, good meat quality, minimal utilization of feed, and high feed conversion rates.

1.2 BREEDING TECHNIQUES

In recent decades, researchers have used standard crossbreeding methods to improve important traits in cattle and poultry. These methods took a long time, had slow genetic progress, and did not work well for breeding traits like disease resistance. Recently, precision molecular breeding has given new ways to speed up the genetic progress of livestock and poultry, to achieve better yield performance, better disease resistance, and less chance of zoonosis exchange, besides better animal welfare (Gao et al. 2023a). This is possible because gene editing technology has developed fast as safety of transgenic techniques is a concern. Gene editing as opposed to transgenic technologies, can change specific genomic spots exactly without adding foreign DNA sequences. It makes it easy to obtain certain genome modifications that occur often at a low rate and produce the same result as natural mutations (Khan et al. 2023). Several gene-edited animals have been approved for sale in different places worldwide, which shows the great potential of gene editing.

1.2.1 Artificial Insemination (AI)

Animal biotechnology relies heavily on artificial insemination (AI), which is revolutionizing reproductive methods and improving our knowledge of animal genetics (Suteky 2021). By facilitating the intentional selection and dissemination of desired genetic features, this approach improves overall production and permits controlled breeding. Utilizing genetically better males whose features may spread quickly throughout a population, AI allows cattle producers to expedite genetic progress (Kefelegn and Alemayehu 2022). Since AI eliminates the necessity for actual mating, it has been shown to be both economical and effective, lowering the possibility of animal harm and disease transmission. It is essential for improving animal productivity overall, genetic progress, and reproductive efficiency. This technique is regarded as a pillar of contemporary animal biotechnology and has had a significant influence on many species. Its capacity to accelerate genetic advancement among herds is one of its main benefits. AI has drawn more attention recently as a potentially effective way to boost agricultural productivity and profitability (Soušková et al. 2021).

Current AI procedures include intrauterine laparoscopic AI using frozen-thawed semen and shallow cervical AI with chilled semen at 15°C. Further research into understudied aspects including the timing of insemination, the physiological state of the female, and semen storage technologies may help improve these approaches (Tomiyasu et al. 2025).

AI has a significant impact on endangered species conservation, in addition to its uses in cattle production. Laser treatments have been used to improve the viability of biological samples since the 1980s. They specifically improve sperm motility and oocyte development by boosting cellular energy through the generation of adenosine triphosphate (ATP). For evaluating the quality of semen, laser-based analytical techniques including laser-induced fluorescence and laser-induced breakdown spectroscopy are being investigated. AI is a very useful and adaptable reproductive method, to sum up. It is a fundamental component of reproductive biotechnology, supporting disease management and species conservation in addition to increasing production and genetic gain (Isnaini et al. 2019; Souskova et al. 2021).

1.2.2 Semen Sexing

The increased rivalry for land between the demands of animal husbandry and a growing human population is a major concern for the global agricultural and livestock sectors. Many approaches have been created over time to deal with this problem, but the majority have been shelved because of their poor performance. Fluorescence-activated cell sorting (FACS) is currently the only popular and reasonably priced method for distinguishing sperm that carry both X and Y chromosomes. This approach, however, is still costly, has little commercial availability, and frequently does not provide semen dosages with sufficient sperm counts for successful insemination. Promising new approaches can be offered by growing proteomic data on X and Y spermatozoa as an alternative to flow cytometry.

According to Quelhas et al. (2021), X and Y sperm have different cell surface proteins, which implies that immunological methods for semen sexing might be successfully created. These approaches are novel and quite promising. Sexed semen is primarily used in dairy heifers to guarantee the birth of female calves for sale as substitutes or for herd expansion. Thawed and sex-sorted frozen semen is rapidly utilized for in vitro fertilization, a technique that is becoming more and more popular, while sexed semen is utilized for targeted matings in elite breeding programs. Furthermore, sexed semen is being used by beef producers to create crossbred female substitutes. It is reported that advances in sperm sexing methods have increased conception rates by around 4%–6% points, or roughly 10% more success rates (Seidel 2014).

1.2.2.1 Applications of Semen Sexing

The production of dairy and beef cattle will receive financial benefits in several ways from sexed semen. Commercial dairy producers can use sexed semen to create beef crossbred sons from the remaining cow population along with substitute daughters from cows that are genetically superior. Seed stock dairy cattle breeders can use sexed semen to make sure all elite dams create a son as it increases the selection response rate. This permits breeders to create bulls for progeny testing; also the number of dams is lower. For that reason, fewer matings are required to make any desired number of daughters. Using sexed semen lowers the progeny testing costs for those bulls.

Semen sorted by the X chromosome allows for single-sex bred-heifer systems. In beef cattle, sexed semen use increases terminal crossbreeding programs' sustainability and efficiency. Fewer females are needed to produce specialized maternal crossbred daughters. This allows more females for the production of highly productive, terminal crossbred sons (Hohenboken 1999). Through the combination of sex-sorted semen and genomic technologies, superior females for replacement needs can be identified. This identification led to the discovery of a dairy cow population from which replacements are not desired. In dairy herds, the use of beef semen greatly increased because of this. Sex-sorted semen is reaching 30% of the entire AI market in North America, but exact numbers are not known. The main purpose of sex-sorted semen is to speed up genetic progress and improve biosecurity by producing replacement animals internally. Additional potential uses are being developed or studied.

1.2.3 Embryo Sexing

Boiler production for males receives more attention than layer strains for females. Early, nonintrusive sex discrimination of chick embryos is necessary, but this is currently not widely practiced. To identify the sex of incubated eggs, the study looks into growth differences based on sex in terms of opacity. Using a grill strain ROSS 308 as a model, eggs were sorted by size, mass, and shell color before incubation. The average output voltage of each egg received was evaluated during the incubation period, using the Embryonic Vital Scope, which consists of LEDs and a Si-photodiode combined with an amplifier that receives the light that passed through the egg. Feather sexing determined the day-old chicks' true gender after hatching (Seidel 2014).

Invasive methods harm an embryo and lower its chance of developing properly. Several such methods are cytological karyotyping, loop-mediated isothermal amplification, finding Y chromosome-specific DNA probes for male chromatin, polymerase chain reaction (PCR) (Xiao et al. 2021), and immunofluorescence (FISH: Fluorescence In Situ Hybridization)). The PCR method is less accurate, sensitive, as well as productive, when compared to invasive methods for embryonic sexing in farm animals. Despite their accuracy, concerns about embryo effectiveness make invasive methods like cytological karyotyping hard to use in practice. This technology can be useful in the dairy and cattle industry, as it helps in producing male and female animals, respectively. It also lowers the cost of many ovulation ETs, besides improving selection accuracy (Bora 2022).

1.2.4 Estrus Synchronization

In the dairy cattle industry, AI breeds most animals. Estrus synchronization is an important tool for reproductive control. Problems with finding estrus, such as high labor costs and errors, can be reduced with estrus synchronization. The basic concept of estrus synchronization is to change the length of the luteal phase of the estrus cycle. Other synchronization programs use a hormone, a few hormones, or a combination of hormones. They have been planned and put into use on commercial farms.

Estrus synchronization manipulates the estrus cycle or induces estrus. This is done to produce estrus in many female animals simultaneously. The estrus cycle refers to the repeated physiological changes from one estrus phase to another in an animal. For cows, the estrus cycle lasts about 21 days, in does, 20 days, and in ewes, 17 days. The hormones progesterone, estrogen, GnRH (Gonadotropin-Releasing Hormone), and prostaglandin take part in estrus synchronization. Depending on the procedures followed, estrus synchronization lowers the breeding period. Instead of females being bred over a typical time, it shortens the period. A specific procedure for cow estrus synchronization using a combination of GnRH with PGF2α (Prostaglandin F_2 alpha) and a progesterone-based synchronization program, was shown to cause changes in follicular development. The estrus synchronization procedure should align with the farmer's desires and management abilities (Arya et al. 2023). When the prostaglandin is administered alone, the cattle injected during late luteal phase exhibit a better estrus response and could be extra fertile than calves inoculated during early luteal phase (Hernandez and Patel 2024).

Recent studies have revealed that the cow is more likely to suffer embryonic and fetal mortality when she is an ovular or an estrus, when her metabolic state is off, when certain nutritional elements are present, or when illnesses arise. While some insemination techniques may affect the survival of the embryo, appropriately timed artificial intelligence has not affected the mortality of the embryo or the fetus in cattle. To maximize conception rates in dairy and beef cattle, future reproductive program improvements must concentrate on increasing fertilization rates and reducing embryonic losses (Santos et al. 2004). A synchronization protocol(7–11 Synch) was developed to synchronize estrus in postpartum beef cows. It synchronized the date of ovulation with the first follicular wave. The effect of the Ovsynch protocol with and without added progesterone on pregnancy rate (PR) in cows was evaluated. These cows had their estrus cycles synchronized before with two doses of PGF2α. During the pre-synchronization period, people did not detect these cows in estrus. A study was conducted in Chihuahua, Mexico, that involved 8,650 Holstein milking cows. The 305-day mature equivalent milk yield was 13,790 kg (Melendez et al. 2006).

Dairy goats were the main subjects of estrus synchronization, to reach the best timing for milk production. But with a growing interest in meat goat production, efforts began to include dairy goat, sheep, and cattle synchronization programs into meat goat management. Methods to achieve synchronization varied. Simple ones included changes in light or social signals (the buck effect), and complex ones combined scheduled hormonal treatments with light changes and the buck effect. Growing consumer and producer interest, together with the import of breeds produced in other countries for rapid development, such the Boer goat, have fuelled the need to develop extremely effective and affordable estrus synchronisation regimens (Whitley and Jackson 2004).

1.2.5 Embryo Transfer (ET)

Embryo transfer is a reproductive biotechnology used in cow breeding to improve genetic performance. A donor cow, often of exceptional genetic quality, is hormonally super ovulated to generate several eggs, which are then fertilised by artificial insemination. About seven days after fertilisation, the resultant embryos are removed non-surgically from the uterus, examined under a microscope, and then either cryopreserved for future use or transplanted into synchronised recipient cows (surrogates).

Compared to natural breeding, this technique enables a single genetically superior cow to give birth to a significantly greater number of offsprings. ET is frequently used in dairy and beef cattle to promote worldwide genetic trade, maintain rare breeds, enhance herd quality, and speed up genetic advancement. The research on ET have evolved over time and these developments have affected its actual use in animal breeding (Betteridge 2003).

A few of the factors that are considered when choosing recipients for successful ETs, include the initial phase which entails paying close attention to the individual's physical state, health, and reproductive characteristics, as well as the herd of origin. The important steps are to prepare, follow up, and handle these females properly. Depending on how these females are managed, the mean pregnancy rates in cattle might range from 55% to 70%. The immunoglobulin G1 assay has the potential to be a valuable tool; however, it can be challenging to interpret early data (Yadav et al. 2025).

For cattle, the commercial practice called "in vitro embryo production" involves large-scale embryo synthesis from slaughterhouse waste plus genetics taken from barren females. A procedure with good results uses ultrasound guidance to aspirate immature oocytes from cattle and buffalo twice a week. Immature oocytes can be removed via laparotomy or laparoscopy from pigs, lambs, and goats. Horses can have immature oocytes removed by ultrasound-guided aspiration. A reduced success rate for in vitro maturation has been reported in pigs and horses. In contrast in vitro maturation is more efficient for cattle, sheep, and goats. Different frozen-thawed semen capacitation procedures have shown the highest success rates for cattle, sheep, and goats (Hasler 1998). The development and use of ET in cattle is a prime example of how professionals should work together for the sake of animal production. Veterinarians, animal scientists, and biologists have shown outstanding collaboration within the International Embryo Transfer Society, which has produced many positive outcomes. The most notable among them is the creation of internationally accepted guidelines for the transnational transfer of embryos (Viana 2019). For both in vivo and in vitro created embryos, evaluation of the possible dangers connected to ET and the effectiveness of washing techniques should be conducted independently (Viana 2024).

1.2.6 In Vitro Fertilization (IVF)

A bull was the first calf born from in vitro fertilization (IVF) in 1981. Since then, fertilization has been done in vitro for cattle, sheep, as well as goats. IVF has produced live offspring in cattle. Some studies indicates in vitro fertilization could help current methods for determining a bull's viability. Further research is required to fully understand about the processes of sperm capping and the ankylose reaction, the maturational competence that oocytes develop, and the unique signals that are present inside the oocyte that allow it to complete maturation, be fertilized, initiate and complete maturation, and eventually result in normal embryo development (García-Roselló and Paramio 2024). After testing spermatozoa and oocyte incubation media, it was shown that neither a medium nor a protein supplement were more effective at encouraging in vitro fertilisation in sheep, pigs, or cattle. There have been reports of successful embryo development in a range of gaseous atmospheres, media and supplements, and culture settings. Results from embryo culture in sealed tubes are comparable to those obtained with medium micro drops submerged in paraffin oil. For bovine embryos, Ham's F-10 medium appears to be the preferred choice; however, a variety of basic culture media perform similarly well for bovine embryos. It does not seem to be very beneficial to supplement medium, although the best bovine serum albumin concentrations are yet unknown. When it comes to fostering embryo growth, a decreased oxygen environment of 5% CO_2, 5% O_2, and 90% N_2 is at least comparable to 5% CO_2 in air and, in certain species, much better (Krisher and Herrick 2024).

With the use of this in vitro method, cow producers—especially those in the dairy sector-have more ways to enhance the distribution of animals with improved genetic value and overcome infertility. Enhancements in Ovum Pick-Up (OPU) and In Vitro Production of embryos (IVP) techniques have led to extensive global commercialization. Compared to other livestock species, the commercial and scientific use of IVP in small ruminants is rather limited. During the preimplantation phase, embryos have to adapt to various microenvironments. Thus, preserving or simulating the in vivo environment in vitro will contribute to enhancing the developing embryo's quality and developmental ability. Both substantial laboratory knowledge and clinical abilities particular to a certain species are needed for the successful clinical implementation of reproductive biotechnology procedures (Wrenzycki 2018).

IVF and early embryonic development of goat oocytes were achieved through experimentation. Large follicles (less than 7mm) of hormone-treated doses and 1–6mm follicles of ovaries from hormone-super stimulated and untreated goats were both sources of oocyte/cumulus complexes. The outcomes achieved with IVM-matured oocytes were notably worse than those with matured in vivo oocytes (68.4% vs. 45.5%, $p< 0.05$). The outcomes of IVM and IVF were better than the control group when LH or FSH was present during egg maturation ($p< 0.05$). Combining the use of mDM for sperm and IVF with IVM in the presence of LH produced the greatest proportion of fertilized oocytes (fertilization rate) (Nguyen et al. 2025).

Farm animal embryos are produced in vitro through three processes: oocyte maturation (IVM), matured oocyte fertilization (IVF), and embryo culture (IVC). Oocytes can be taken from animals that are still alive (by laparotomy, laparoscopy, or ovum pick-up) or from animals that have been killed (by puncture or sectioning). IVM is the process in which immature oocytes are typically separated and matured. Zygotes are created by the culturing of mature oocytes with sperm (IVF). Intracytoplasmic sperm injection is utilized in cases of poor fertilization (pigs, horses). Usually, the zygotes are cultivated (IVC) until they reach the blastocyst and morula phases. These embryos may be vitrified, frozen, or transferred to recipients. To date, cattle, sheep, goat, pig, and horse embryos have been transferred successfully to obtain offspring (Duszewska et al. 2012).

1.2.7 Cryopreservation

Cryopreservation in cattle is the method of storing gametes (sperm and oocytes) or embryos at extremely low temperatures (often in liquid nitrogen at 196 °C) in order to retain their viability for long-term storage and usage. For some farm animals, sperm freezing (cryopreservation of semen) brings many advantages, and its importance became clearer in recent years. It is a very important process in the development of technologies for assisted reproduction. In many studies, the addition of certain substances improved sperm quality and ability to fertilize after thawing. In addition the possible use of molecular biomarkers to predict sperm tolerance to freezing is also reported.

Mammalian embryo cryopreservation has several important processes. In cattle, compacted morulae and blastocysts have good survival, but oocytes plus early development stages do not survive well in vivo or in vitro. In pigs, blastocysts have better survival compared to other development phases. The special characteristics of horse embryos cause problems for successful freezing (Massip 2001). Melatonin (MT) is an antioxidant with practical uses in many fields. To protect against damage from oxidative stress in frozen-thawed sperm, MT is often used in sperm cryopreservation because it can remove free radicals and increase cellular antioxidant defenses. To protect membrane along with DNA integrity involves direct or indirect removal of free radicals, stopping lipid peroxidation, lowering oxidative stress, increasing post-thaw antioxidant and enzymatic functions to keep mitochondrial functions and activity, and controlling ATP production and use to retain sperm quality and motility. Furthermore, it has been observed that MT prevents sperm apoptosis, perhaps via increasing sperm viability and adjusting mRNA transcript abundances (Ofosu et al. 2021).

The first successfully produced progeny from frozen-thawed kernels were later observed in the various lab animals, including mice. With cryopreserved cow, sheep, goat, and horse embryos, favorable outcomes have been achieved thus far in terms of pregnancy rates (PRs); however, no success with swine embryos has been documented. It has been demonstrated that the species, developmental stage, and initial embryo quality all affect post-thaw embryo survival. The term "equilibrium" cryopreservation refers to the freezing methods that are most commonly employed in scientific studies and by businesses.

Recent studies have analyzed that a "nonequilibrium" approach may also be used to freeze embryos. Through the use of extremely intense vitrification medium, the embryo is dehydrated to room temperature before being rapidly frozen to prevent ice formation and enable a solution to transition from a liquid to a glassy state. This process is known as fast freezing by vitrification. Solutions for vitrification are mixtures of propylene glycol, glycerol, and sucrose. Fifty percent PRs with the bovine blastocyst have been recorded using this procedure (Smith 2025).

1.3 GENETIC BASIS OF FARM BIOTECHNOLOGY

Population growth and the necessity for social development created intense interest in genetic improvement of farm animals within agricultural research. Traditional breeding methods—including directed selection breeding—failed to meet requirements for sustainable agricultural production in recent years. Researchers concentrate on quicker and more productive approaches to genetic enhancement. Gene editing, a genetic engineering method, directs at a particular location in an organism's genome and introduces precise changes, for example, insertions, deletions, or mutations, using nonhomologous end joining or homologous recombination (Salsman and Dellaire 2017).

The advent of various potent molecular tools, including as the Clustered Regularly Interspaced Short Palindromic Repeats—CRISPR associated protein 9 (CRISPR/Cas9) system, zinc finger nucleases (ZFNs), and transcription activator-like effector nucleases (TALENs), marked the beginning of the age of gene editing. The development of medicine and the biological sciences has been revolutionized by these instruments. The genetic enhancement of farm animals has been greatly influenced by gene editing techniques. Pigs that lack the myostatin (MSTN) gene, which is generated using ZFNs, for example, have double muscling and improved growth characteristics. In a similar vein, TALENs have been used to produce hornless cattle, enhancing the safety and well-being of animals. The production of hens immune to the avian leucosis virus, a serious illness affecting poultry, has been made possible via CRISPR/Cas9 technology. These tools all provide different mechanisms: TALENs employ transcription activator-like effectors for targeted gene binding, ZFNs use designed zinc-finger domains to recognize DNA sequences, and CRISPR/Cas9 has become the most popular method because of its ease of use, accuracy, and efficiency in genome editing (Bhardwaj and Nain 2021).

These kinds of gene editing methods (Figure 1.1) have been widely used in cattle and poultry (Lee et al. 2020a), significantly advancing molecular breeding in these animals.

Generally speaking, the process of creating livestock species with altered genes involves first modifying the target gene in cultured cells, and then creating the animals via embryonic technologies such somatic cell nuclear transfer (SCNT) and human-handled cloning. Furthermore, simpler techniques like microinjection (MI) and electroporation following in vitro fertilization (gene editing via electroporation, or GEEP) are used to create gene-edited farm animals. Fowl requires different techniques than mammals to produce gene-edited fowl because of its distinct reproductive traits. The ability to precisely modify target genes in chickens has improved recently due to the development of in vitro growth of chicken primordial germ cells (PGCs) (Mathan et al. 2023), chimaera production technology based on PGCs, and more effective gene editing technology. TALENs, CRISPR, and homologous recombination have all been effectively used in chickens so far (Khwatenge and Nahashon 2021). Though PGC is an effective tool for precise genome editing, only chicken PGC can be reliably cultivated in vitro, and PGC culture has not yet been developed for other poultry species. In addition to the PGC method, other techniques for creating gene-edited birds include adenovirus-mediated blastoderm insertion (Lee et al. 2020b), vascular injection of plasmids (Liu et al. 2019), and sperm transfection-assisted gene editing (Zhang et al. 2022). However, these techniques still need to be refined, and fresh approaches must be created for gene editing in poultry in a way that is more straightforward, effective, and adaptable. Poultry molecular breeding is expected to have a promising future in disease resistance breeding, egg bioreactor, and performance improvement. Over the last 10 years, there has been a rapid advancement in the oversight, assessment, and control of animals that have undergone gene editing, and a number of farm animal-derived goods have been authorized for commercialization. The FDA

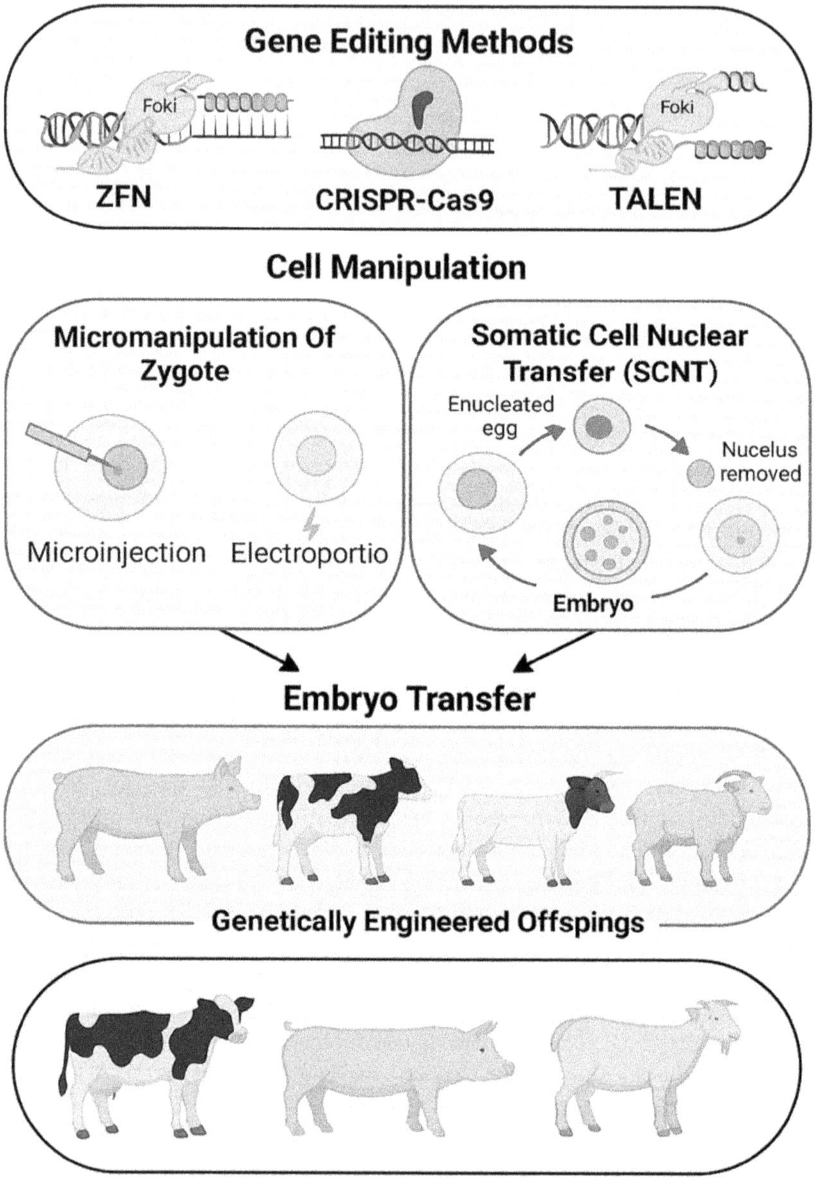

FIGURE 1.1 Gene-edited farm animals by gene editing tools like ZFNs, CRISPR/Cas9, and TALENs.

authorized a recombinant anti-thrombin medication made from transgenic goat milk in 2009, a medication expressing genetically engineered chicken in 2015, the Aqu Advantage, genetically modified salmon in 2015—the first edible gene-edited animal, and GalSafe pigs devoid of α-galactose for food and medicinal applications in 2020. These approvals suggest that more gene-edited animals will be identified and that the market will become more diverse (Turnbull et al. 2021).

1.4 MULTIGENE EDITING IN FARM ANIMALS

Traditional livestock breeding isa slow process, as their breeding cycles are longer; it takes many years especially in large animals. New gene editing techniques can help address the above-mentioned issue, as these techniques are fast and precise, and they can manipulate multiple genes at once, which is beneficial for breed ersto have desired traits in short time. For instance, Zhu et al. (2022) used CRISPR and SCNT to produce pigs with two genes CD163 and APN turned off at same time. These pigs have control over Porcine Reproductive and Respiratory Syndrome Virus (PRRSV) and Transmissible Gastroenteritis Virus (TGEV), meaning they are resistant to both viruses as CD163 and APN genes act as receptors for these two viruses respectively. PRRSV causes reproductive failure in sows and respiratory disease in pigs whereas TGEV causes severe diarrhea and dehydration in piglets. Thus, studies on these two viruses is very crucial for swine health management and vaccine development. The pigs produced have same meat quality and reproduction rate as normal pigs.

Triple gene editing is another method that is done in one step, which uses HA3A-BE3-Y130F base editor that is injected into pig zygote. With this, three genes can be changed together. They are CD163, MSTN, and IGF2. Pigs with triple gene editing showed effective decreases in CD163 and MSTN expression levels and increase in IGF2 expression. According to Gao et al. (2023b), these modifications took place as planned, producing pigs that showed increased disease resistance and growth performance. In a similar work, Popova et al. (2023) used gene editing to enhance goats' meat output and cashmere productivity. It was done by injecting Cas9 mRNA and single-guide RNAs into goat zygotes. The genes that were targeted were FGF5 (linked to hair growth) and MSTN (for regulating muscle growth). There were 98 goat offspring from the altered embryos. About 10% of the goats showed effective knockouts of both genes, with the gene knockout efficiency for FGF5 being 21% and MSTN being 15%.

For the implementation of molecular breeding in livestock, these studies (Table 1.1) are important both theoretically and practically. The off-target effect has grown to be a significant issue for researchers in recent years due to the advancement of effective gene editing technology. In order to create a broadly applicable off-target activity assay (NT-SEQ) for a variety of gene editing tools, including base editing tools, researchers have integrated SCNT with whole genome sequencing technology (Gao et al. 2023a). The development of this off-target activity detection technique will direct ongoing gene editing tool optimization and expand the tools' potential uses. In addition to offering technical assistance for resolving global food security issues and guaranteeing food supply, ongoing improvement of the efficiency and precision of gene editing technology may significantly encourage the widespread use of gene-edited large animal models.

1.5 GENETIC TECHNOLOGIES USED IN BREEDING

Over the past 50 years, quantitative genetic innovations have revolutionized animal agriculture by creating advanced techniques for accurately calculating an animal's genetic value, also known as estimated breeding value, or EBV (Salamone et al. 2018). These systems, particularly Best Linear Unbiased Prediction (BLUP), have been used in all major cattle species. Thanks to new molecular genetics techniques, it is possible

TABLE 1.1
Genetic Modifications in Animals

Year	Milestone	Strategy	References
1985	First transgenic sheep and pigs	Microinjection of DNA into one pronucleus of a zygote	Hammer et al. (1986)
1986	Embryonic cloning of sheep	Nuclear transfer using embryonic cells as donor cells	Willadsen (1986)
1997	Transgenic sheep produced by nuclear transfer	Random integration of the construct	Schnieke et al. (1997)
1998	Transgenic cattle produced from fetal fibroblasts	Random integration of the construct	Hasler (1998)
2000	Gene targeting in sheep	Gene replacement and nuclear transfer	McCreath et al. (2000)
2003	Transgenic pigs via lentiviral injection	Gene transfer into zygotes via lentiviruses	Hofmann et al. (2003)
2011	Conditional transgene expression in pigs	Pronuclear DNA injection and crossbreeding	Kues and Niemann (2011)
2015	Cloning of sheep with somatic donor cells	Nuclear transfer using adult somatic donor cells	Wilmut et al. (2015)

to have better understanding of the genetic composition of animals and the function of genes that control complex production features. The use of this genetic information may lead to a paradigm change in animal breeding and production, as discussed below.

1.5.1 DNA and DNA-Based Selection

The development of molecular and related analytical genetic techniques over the past few decades, such as DNA libraries, physical and linkage maps, and, more recently, whole genome sequences, has allowed us to identify any gene in the genome. The first application of this technique was the development of DNA markers for identification and parentage determination. Quantitative trait loci (QTL) linked to genetic variation in production traits and disease were then monitored using it. At the lower end of the DNA technology spectrum, a low-cost DNA-based parentage assignment could fill a critical information gap in BLUP-based evaluation systems (Dilger et al. 2022). For breeding programs and genetic evaluation to be effective, access to complete or partial pedigree data is necessary. Most sheep breeding systems and large beef breeding systems do not currently have access to pedigree data, and BLUP-based rating of breeding replacements is often limited to phenotypic records of candidates that are being selected. The ability to track commercial animals from the point of harvest (such as shearing and slaughter) to the commercial tier sire families is another benefit of incredibly inexpensive DNA identification assignment. This will increase the profitability of the current generation by enabling the removal of underperforming sires and giving the seed stock industry access to more genetic data.

Single genes that cause hereditary diseases and genes that have a major impact on production features have now been identified using molecular genetic techniques:

Online Mendelian Inheritance in Animals (OMIA) lists 126 diseases or characteristics for which the underlying mutation has been identified at the DNA level. In some cases, DNA testing can definitively identify whether a potential breeding candidate has zero, one, or two copies of a specific mutation. Many QTL have been found for each of the major livestock species (swine: Talebi et al. 2022; cattle: Diez-Tascon et al. 2001; Ron and Weller 2007; sheep: Diez-Tascon et al.2001), making it a widely used testing technique. According to experimental findings, the use of QTL in commercial breeding programs supported by MAS has been delayed. Until a large panel of these QTL has been reduced to the specific mutations causing output variations, allowing direct selection across multiple loci, it is unclear whether MAS will be helpful in the cattle sector generally. The availability of comparative gene function data across mammalian species and cattle genome data has already aided in the development of direct gene markers for polygenic traits (Giuffra et al. 2019). To be commercially more attractive than conventional techniques for predicting genetic quality, selection would still need to find enough distinct gene markers for each attribute. It is anticipated that DNA-based data will be obtained at prices that enable incredibly high-throughput analysis (human genomic analysis arrays currently cover over 100,000 markers per scan), possibly in a disposable format at the collection location (farm, saleyard, or abattoir). As a result, it will soon be feasible to analyze multiple QTL loci that are known to affect production traits at once. With a high degree of reliability, the application of such technology offers notable improvements in the early (possibly embryonic) genetic merit prediction. DNA marker-based selection will be most commercially beneficial when used for traits that are expensive and difficult to measure (e.g., disease resistance), expressed late in life (e.g., lifetime productivity or reproductive performance), or expressed in only one sex (e.g., maternal or reproductive performance). The technique enables the selection of superior or inferior individuals at very young ages and the targeted introduction of novel allelic variations with significant effects in new target populations. When paired with faster reproduction techniques, this technology will be most effective.

1.5.2 Gene Expression Profiles

The conventional emphasis on individual genes and proteins has given the way to a more integrated approach that enables the simultaneous evaluation of multiple genes, their products, functional activity, and interactions thanks to developments in mammalian proteomics and genomics. High-throughput technology advancements have made this shift easier, with microarrays becoming one of the most popular instruments for extensive gene activity screening. On delicate membranes or glass slides, thousands of gene samples are systematically arranged in a high-density pattern to form microarrays. Up to 30,000 identified mammalian genes can be found in some of the most sophisticated microarrays. These ever-evolving technologies make it possible to investigate intricate biological systems more thoroughly and precisely.

The activity levels of individual genes within a particular tissue of an organism at a given time can be determined with the help of microarrays. The development of pharmacological and genetic strategies to modulate or regulate metabolic pathways is made easier by this capability, which improves our understanding of the genetic basis of these processes. These "gene expression phenotypes" can be used as indirect selection parameters to evaluate physiological status and are high-resolution indicators for complex traits. They are also very useful in studies that focus on gene function. Even at early developmental stages, these "advanced phenotypes" can be gathered for a variety of functional pathways. Traditional genetic analysis is still necessary to assess these phenotypes in order to find gene variation and ascertain how it correlates with traits that are important for breeding goals (Kumar et al. 2023).

1.5.3 DNA Diagnostics of Pathogens

The ability of biotechnology to create targets based on DNA or RNA for improved diagnostic testing is well established (Thavarajah et al. 2021). Additionally, diagnostic technology will feature high-capacity, real-time animal screening for QA standards in slaughterhouses and sale yards, as well as dependable and user-friendly test kits for use on farms. This type of technology will continue to advance to deliver quick, inexpensive, sensitive, and extremely accurate disease information for animal applications with relatively unconstrained growth and extension. In addition to the DNA-based techniques described above, supply chain operations can use high-value data gathered at the commercial "take-off" point.

1.5.4 Genetic Modification

The alteration in an organism's native DNA code is referred to as genetic modification or genetic engineering. This involves duplicating or removing genes that are already present, as well as inserting alien or unrelated DNA. Integrating high-value data gathered at the commercial "take-off" point. Numerous technologies are prepared to implement these adjustments. The majority of them use vectors. In addition, the vector aids in the insertion of DNA into cells (Bihon and Assefa 2019). To modify germ-line cells, some of these are employed. As a result, the alterations are passed down to the children in the following generation. Other technologies are only intended to modify somatic cells. Furthermore, for a variety of production objectives, the most appealing method for producing transgenic animals in cattle is using ART (Shah and Chauhan 2019).

1.5.5 Somatic Gene Transfer (Gene Therapy)

Human hereditary disorders have been treated via somatic gene transfer in a few, rare cases (Skrzyszowska and Samiec 2021). Gene transfer can be done in two ways: in vivo, which includes delivering the patient gene vectors directly, and ex vivo, which entails genetically altering cells in a lab before injecting them into the body. Numerous viral and physical delivery techniques, with vectors such as retrovirus, adenovirus, lentivirus, and adeno-associated virus, as well as electroporation, MI, particle bombardment, and lipofection, are being researched. Insert capacity, integration, integration site, transgenic expression length and level, ease of delivery, tissue specificity, and transfection efficiency are only a few of the advantages and disadvantages of the various approaches. Fortunately, because of the high degree of genome similarity between humans, model animal systems, and cattle, the agriculture sector may benefit from the far greater investment in medical genomics through comparative functional and structural biology.

1.5.6 DNA Vaccination

Compared to conventional vaccination techniques, DNA vaccines have a number of benefits, such as stronger immunological memory, improved antigen presentation, and improved immune response profiles (Pagliari et al. 2023). Since DNA vaccines do not need to be kept in a cold chain, they are thought to be safer, more affordable, simpler to make, and more stable than live, inactivated, or protein-based vaccines. Usually, a bacterial plasmid with a foreign gene and a promoter makes up a DNA vaccine. After being cloned into a plasmid, this gene is inserted into mammalian tissue, usually deep muscle. The target immunogen, a protein that triggers an immune response, is created when local cells absorb the plasmid DNA and transcribe and translate the encoded gene. The immune system can identify and protect against recurrent exposure to the pathogen thanks to this reaction.

In addition to being quick to develop and produce, DNA vaccines have long-term stability and fewer adverse effects. They are a promising platform for veterinary and human applications due to their versatility and ease of production. DNA vaccines have potential applications in animal husbandry beyond their medical use, such as controlling reproductive cycles and interfering with negative feedback systems related to body composition and growth. As a result, DNA vaccines are becoming recognized as a useful and cutting-edge instrument for livestock management and healthcare.

1.5.7 Gene Silencing

RNAi is a process of expression-specific posttranscriptional gene silencing in which a targeted mRNA is bound by tiny RNA molecules, causing the mRNA to begin to degrade (Gandhi 2023). As a result, up to 90% less of the protein transcribed by the mRNA is produced. This method is highly helpful for studying the function of mammalian genes, for preventing the expression of genes that cause disease, and it has the potential to be used to externally downregulate particular gene products in cattle.

1.5.8 Germ-Line Genetic Modification

In their study of the effects of GM on the animal industry, Pasitka et al. (2023) emphasized the use of pronuclear injection as one of the first techniques for introducing foreign DNA into animals. Using this method, DNA is directly inserted into embryos or embryonic stem cells. Pronuclear injection was considered revolutionary once, but the method is today regarded as extremely inefficient since the transgene is effectively incorporated into just 3%–5% of the resultant animals. Additionally, it frequently results in random integration sites within the genome, mosaicism, and uncontrolled copy number insertion. These technological issues and the scarcity of germ-line stem cells have made the technique less popular.

SCNT and in vitro genetic engineering of somatic cells are the current standards for livestock GM. This method allows for precise DNA manipulation by genetically modifying somatic cells and cultivating them under controlled conditions. Following nuclear transfer, these altered cells provide better control and greater consistency when creating genetically modified animals. However, the effectiveness of the nuclear transfer process as well as other technical issues continue to limit this technique's overall success. Notwithstanding these drawbacks, the technique has a lot of potential because transgenic livestock may have advantages. More accurate genetic changes are made possible by homologous recombination, which permits targeted integration at particular genomic locations. Consequently, the offspring of these genetically modified animals can acquire desired characteristics without the negative consequences that are frequently linked to nuclear transfer (Bishop and Van Eenennaam 2020). Additionally, it is anticipated that developments in functional genomics will make it easier to create a large number of tissue-specific promoters that can react to particular stimuli, improving the accuracy and usefulness of GM in livestock.

1.6 CONCLUSION AND FUTURE PROSPECTS

The cattle industry's long-term financial survival mostly rests on halting the present trade downturn. Long-term productivity increases of 1.5%–2.5% are necessary to counteract growing operating costs in light of the worldwide trend of price compression across agricultural systems. Promising ways to increase the value of cattle output are provided by biotechnology. It makes it possible to create goods with distinctive, marketable features and helps to increase the efficiency and cost-effectiveness of production processes, which lowers the cost per unit of output in the end.

Reproductive technologies are essential because they provide breeders control, enabling more rigorous selection and shorter generational gaps. In parallel, by making it possible to identify and choose desirable features, DNA-based selection techniques greatly speed up genetic gains. Targeting attributes like disease resistance, sex-restricted traits, or traits with limited expression that would otherwise be costly, time-consuming, or difficult to evaluate is now possible thanks to these techniques. Biotechnology offers new possibilities for animal breeding that are beyond the potential of conventional techniques, especially when combined with genetic engineering. Conventional breeding lays the groundwork, but it is doubtful that consumers will oppose the integration of reproductive and genetic technology, particularly if it is in line with accepted genetic evaluation procedures. This compatibility could help the industry embrace it more quickly.

These developments also pave the way for the creation of valuable animal lines that are not limited by present biological limitations. However, resolving important societal issues including social justice, environmental sustainability, animal welfare, and public acceptance is essential to the effective use of such technology. It becomes more and more possible to incorporate cutting-edge biotechnologies into conventional cattle production once these issues are sufficiently resolved.

REFERENCES

Arya, D., R. Goswami, and M. Sharma. 2023. Estrus synchronization in cattle, sheep and goat. *Multidiscip. Rev.* 61: 2023001–2023001.

Bai, X., and G.S. Plastow. 2022. Breeding for disease resilience: Opportunities to manage polymicrobial challenge and improve commercial performance in the pig industry. *CABI Agric. Biosci.* 3(1): 1–17.

Betteridge, K.J. 2003. A history of farm animal embryo transfer and some associated techniques. *Anim. Reprod. Sci.* 793: 203–244.

Bhardwaj, A., and V. Nain. 2021. TALENs—an indispensable tool in the era of CRISPR: A mini review. *J. Genet. Eng. Biotechnol.* 19(1): 1–10.

Bihon Asfaw, A., and A. Assefa. 2019. Animal transgenesis technology: A review. *Cogent Food Agric.* 51: 1686802.

Bishop, T.F., and A.L. Van Eenennaam. 2020. Genome editing approaches to augment livestock breeding programs. *J. Exp. Biol.* 223: jeb207159.

Bora, S.K. 2022. Embryo sexing methods in bovine and its application in animal breed. *J. Anim. Reprod. Biotechnol.* 372: 80–86.

Diez-Tascon, C., Y. Bayon, J.J. Arranz, F. De La Fuente, and F. San Primitivo. 2001. Mapping quantitative trait loci for milk production traits on ovine chromosome6. *J. Dairy Res.* 68: 389–397.

Dilger, A.C., X. Chen, L.T. Honegger, B.M. Marron, and J.E. Beever. 2022. The potential for gene-editing to increase muscle growth in pigs: Experiences with editing myostatin. *CABI Agric. Biosci.* 3: 1–14.

Duszewska, A., Ł. Rąpała, P. Trzeciak, S. Dąbrowski, and A. Piliszek. 2012. Obtaining farm animal embryos in vitro. *J. Anim. Feed Sci.* 212: 217–233.

Foley, J.A., N. Ramankutty, K.A. Brauman, E.S. Cassidy, J.S. Gerber, M. Johnston, and D.P. Zaks. 2011. Solutions for a cultivated planet. *Nature.* 478: 337–342.

Gandhi, N.V. 2023. *Biotechnology in Medicine*(p. 105). Weser Books.

Gao, F., N. Hou, X. Du, Y. Wang, J. Zhao, and S. Wu. 2023a. Molecular breeding of farm animals through gene editing. *Nat. Sci. Open.* 2: 20220066.

Gao, F., P. Li, Y. Yin, X. Du, G. Cao, S. Wu, and Y. Zhao. 2023b. Molecular breeding of livestock for disease resistance. *Virology.* 587: 109862.

García-Roselló, E., and M.T. Paramio. 2024. Advances in in vitro fertilization techniques in ruminants: Current challenges and future prospects. *Reprod. Fertil. Dev.* 36(2): 123–136.

Gardon, J.C., and K. Satue (Eds.). 2020. *Biotechnologies Applied to Animal Reproduction: Current Trends and Practical Applications for Reproductive Management.* CRC Press.

Giuffra, E., C.K. Tuggle, and FAANG Consortium. 2019. Functional annotation of animal genomes (FAANG): Current achievements and roadmap. *Annu. Rev. Anim. Biosci.* 7: 65–88.

Godfray, H.C.J., J.R. Beddington, I.R. Crute, L. Haddad, D. Lawrence, J.F. Muir, and C. Toulmin. 2010. Food security: The challenge of feeding 9 billion people. *Science.* 327: 812–818.

Gullo, G., G. Basile, G.Cucinella, M.E. Greco, A.Perino, V. Chiantera, and S. Marinelli. 2023. Fresh vs. frozen embryo transfer in assisted reproductive techniques: A single center retrospective cohort study and ethical-legal implications. *Eur. Rev. Med. Pharmacol. Sci.* 27(14): 14.

Hammer, R.E., V.G. Pursel, C.E. Rexroad, R.J. Wall, D.J. Bolt, K.M. Ebert, R.D. Palmiter, and R.L. Brinster. 1986. Production of transgenic rabbits, sheep and pigs by microinjection. *Nature.* 315: 680–683.

Hasler, J.F. 1998. The current status of oocyte recovery, in vitro embryo production, and embryo transfer in domestic animals, with an emphasis on the bovine. *J. Anim. Sci.* 76(Suppl. 3): 52–74.

Hernandez, J.L., and S. Patel. 2024. Advances in estrus synchronization protocols for postpartum beef and dairy cattle: An updated review. *J. Anim. Sci.* 102(4): 945–961.

Hofmann, A., B. Kessler, S. Ewerling, M. Weppert, B. Vogg, H. Ludwig, and A. Pfeifer. 2003. Efficient transgenesis in farm animals by lentiviral vectors. *EMBO Reports*. 4: 1054–1058.

Hohenboken, W. 1999. Applications of sexed semen in cattle production. *Theriogenology*. 52: 1421–1433.

Isnaini, N., S. Wahjuningsih, A. Ma'ruf, and D. Witayanto. 2019. Effects of age and breed on semen quality of beef bull sires in an Indonesian artificial insemination center. *Livest. Res. Rural Dev*. 31: 5.

Kefelegn, S., and L. Alemayehu. 2022. Breeding soundness of semen producing bulls, their semen quality and interrelationship of functional tests with spermatozoa motility percentage. *J. Vet. Med. Anim. Health*. 14: 1–11.

Khalil, A.M. 2025. Traditional strategies and cutting-edge technologies used for animal breeding: A comprehensive review. *Sustain Food Connect*. 1: 1–10.

Khan, A., N. Paneerselvam, and B.R. Lawson. 2023. Antiretrovirals to CCR5 CRISPR/Cas9 gene editing: A paradigm shift chasing an HIV cure. *Clin. Immunol.* 256: 109741.

Khwatenge, C.N., and S.N. Nahashon. 2021. Recent advances in the application of CRISPR/Cas9 gene editing system in poultry species. *Front. Genet.* 12: 627714.

Krisher, R.L., and J.R. Herrick. 2024. Bovine embryo production in vitro: Evolution of culture media and commercial perspectives. *Anim. Reprod.* 21(3): e20240051.

Kues, W.A., and H .Niemann. 2011. Advances in farm animal transgenesis. *Prev. Vet. Med.* 102: 146–156.

Kumar, A., G. Kumar Gaur, M. Panigrahi, M.A. VN, B. Priya, J.P. Gupta, and R.K. Gandham. 2023. Comparative gene expression profile in circulating PBMCs of Bos indicus and crossbred cattle to understand disease tolerance mechanism. *Anim. Biotechnol.* 34: 1594–1602.

Lee, J., D.H. Kim, and K. Lee. 2020a. Current approaches and applications in avian genome editing. *Int. J. Mol. Sci.* 21: 3937.

Lee, K., K. Uh, and K.F arrell. 2020b. Current progress of genome editing in livestock. *Theriogenology*. 150: 229–235.

Liu, M., L. Zhang, G. Marsboom, A. Jambusaria, S. Xiong, P.T. Toth, and A.B. Malik. 2019. Sox17 is required for endothelial regeneration following inflammation-induced vascular injury. *Nat. Commun.* 10: 2126.

Massip, A. 2001. Cryopreservation of embryos of farm animals. *Reprod. Domest. Anim.* 36: 49–55.

Mathan, Z., G. Jin, K. Zuo, M. Habib, Y. Zhang, and B. Li. 2023. Formation, application, and significance of chicken primordial germ cells: A review. *Animals*. 13: 1096.

McCreath, K.J., J. Howcroft, K. Campbell, A. Colman, A. Schnieke, and I. Kind. 2000. Production of genetically modified sheep by gene targeting in somatic cells. *Nature*. 405: 1066–1069.

Melendez, P., G. Gonzalez, E. Aguilar, O. Loera, C. Risco, and L. Archbald. 2006. Comparison of two estrus-synchronization protocols and timed artificial insemination in dairy cattle. *J. Dairy Sci.* 89: 4567–4572.

Miller, E.R., and M.E. Fowler. 2014. *Fowler's Zoo and Wild Animal Medicine*, Vol. 8. Elsevier Health Sciences.

Miller, S. 2022. Genomic selection in beef cattle creates additional opportunities for embryo technologies to meet industry needs. *Reprod. Fertil. Dev*. 35(2): 98–105.

Nguyen, V.K., H.T.T. Vu, A.T. Hoang, Y.T.K. Pham, D.V. Le, Q.X. Huu, L.A.T. Nguyen, H.L.T. Nguyen, and L.D. Pham. 2025. Optimization of the in vitro fertilization system in Cportunities fo. *Open J. Anim. Sci.* 15(1): 39–50.

Niemann, H., and B. Seamark. 2018. The evolution of farm animal biotechnology. In: H. Niemann and C. Wrenzycki (Eds.), *Animal Biotechnology 1: Reproductive Biotechnologies*(pp. 1–26). Springer.

Niemann, H., and W.A. Kues. 2007. Transgenic farm animals: An update. *Reprod. Fertil. Dev.* 19: 762–770.

Ofosu, J., I.H. Qazi, Y. Fang, and G. Zhou. 2021. Use of melatonin in sperm cryopreservation of farm animals: A brief review. *Anim. Reprod. Sci.* 233: 106850.

Pagliari, S., B. Dema, A. Sanchez-Martinez, G.M.Z. Flores, and C.S. Rollier. 2023. DNA vaccines: History, molecular mechanisms and future perspectives. *J. Mol. Biol.* 435: 168297.

Pasitka, L., M. Cohen, A. Ehrlich, B. Gildor, E. Reuveni, M. Ayyash, and Y. Nahmias. 2023. Spontaneous immortalization of chicken fibroblasts generates stable, high-yield cell lines for serum-free production of cultured meat. *Nat. Food.* 4: 35–50.

Popova, J., V. Bets, and E. Kozhevnikova. 2023. Perspectives in genome-editing techniques for livestock. *J. Anim.* 13: 2580.

Quelhas, J., J. Santiago, B. Matos, A. Rocha, G. Lopes, and M. Fardilha. 2021. Bovine semen sexing: Sperm membrane proteomics as candidates for immunological selection of X- and Y-chromosome-bearing sperm. *Vet. Med. Sci.* 75: 1633–1641.

Ron, M., and J.I. Weller. 2007. From QTL to QTN identification in livestock—winning by points rather than knock-out: A review. *Anim. Genet.* 38: 429–439.

Salsman, J., and G. Dellaire. 2017. Precision genome editing in the CRISPR era. *Biochem. CellBiol.* 95: 187–201.

Santos, J., W. Thatcher, R. Chebel, R. Cerri, and K. Galvao. 2004. The effect of embryonic death rates in cattle on the efficacy of estrus synchronization programs. *Anim. Reprod. Sci.* 82: 513–535.

Schnieke, A.E., A.J. Kind, W.A. Ritchie, K. Mycock, A.R. Scott, M. Ritchie, and K.H. Campbell. 1997. Human factor IX transgenic sheep produced by transfer of nuclei from transfected fetal fibroblasts. *Science.* 278: 2130–2133.

Seidel, G. 2014. Update on sexed semen technology in cattle. *J. Anim.* 8: 160–164.

Shah, S.M., and M.S. Chauhan. 2019. Modern biotechnological tools for enhancing reproductive efficiency in livestock. *Indian J. Genet. Plant Breed.* 79: 241–249.

Skrzyszowska, M., and M. Samiec. 2021. Generating cloned goats by somatic cell nuclear transfer—Molecular determinants and application to transgenics and biomedicine. *Int. J. Mol. Sci.* 22: 7490.

Smith, K.L. 2025. Advancements in cryopreservation of domestic animal embryos. *Domest. Anim. Endocrinol.* 92: 106938.

Soušková, K., R. Filipčík, and Z. Rečková. 2021. Evaluation of stallion sperm concentration by two different methods and its influence on sperm motility assessment. *Acta Fytotechn. Zootechn.* 24(4): 293–296.

Suteky, T. 2021. Factors affecting adoption of beef cattle artificial insemination (AI) in smallholder farms in Bengkulu Province, Indonesia. *Proc. Int. Sem. Promot. Local Resour. Sustain. Agric. Dev. (ISPLRSAD2020).* 13: 194–198.

Talebi, R., M.R. Ghaffari, M. Zeinalabedini, R. Abdoli, and M. Mardi. 2022. Genetic basis of muscle-related traits in sheep: A review. *Anim. Genet.* 53: 723–739.

Thavarajah, W., L.M. Hertz, D.Z. Bushhouse, C.M. Archuleta, and J.B. Lucks. 2021. RNA engineering for public health: Innovations in RNA-based diagnostics and therapeutics. *Annu. Rev. Chem. Biomol. Eng.* 12: 263–286.

Todd, D.L. 2013. Implementation of genomic selection in UK beef and sheep breeding. Doctoral dissertation, University of Edinburgh.

Tomiyasu, J., N. Matsumoto, H. Katsushima, and Y. Yanagawa. 2025. Association between back scent gland development and reproductive status in male brown bears (*Ursus arctos). J. Exp. Zool. Part A: Ecol. Integr. Physiol.* 343(2): 115–123.

Turnbull, C., M. Lillemo, and T.A. Hvoslef-Eide. 2021. Global regulation of genetically modified crops amid the gene-edited crop boom: A review. *Front. Plant Sci.* 12: 630396.

Viana, J. 2019. 2018 statistics of embryo production and transfer in domestic farm animals. *Embryo Technol. Newslett.* 36: 17–26.

Viana, J.H.M. 2024. Development of the world farm animal embryo industry over the past 30 years: Sanitary perspectives. *Theriogenology*. 230: 151–156.

Whitley, N.C., and D. Jackson. 2004. An update on estrus synchronization in goats: A minor species. *J. Anim. Sci.* 82(suppl_13): E270–E276.

Willadsen, S.M. 1986. Nuclear transplantation in sheep embryos. *Nature*. 320(6057): 63–65.

Wilmut, I., Y. Bai, and J. Taylor. 2015. Somatic cell nuclear transfer: Origins, the present position and future opportunities. *Philos. Trans. R. Soc. B*. 370(1680): 20140366.

Wrenzycki, C. 2018. In vitro production of farm animal embryos. *Anim. Biotechnol.* 1: 269–304.

Xiao, Y., F. Sosa, L.R. de Armas, L. Pan, and P.J. Hansen. 2021. An improved method for specific-target preamplification PCR analysis of single blastocysts useful for embryo sexing and high-throughput gene expression analysis. *J. Dairy Sci.* 1043: 3722–3735.

Yadav, J.N., H.K. Singh, V. Yadav, and R.P. Diwakar. 2025. Artificial insemination and embryo transfer: Emerging technologies in the livestock industry. *Int. J. Environ. Agric. Biotechnol.* 10(2): 119–128.

Yánez-Ortiz, I., J. Catalán, J.E. Rodríguez-Gil, J. Miró, and M. Yeste. 2022. Advances in sperm cryopreservation in farm animals: Cattle, horse, pig and sheep. *Anim. Reprod. Sci.* 246: 106904.

Zhang, X., H. Jin, X. Huang, B. Chaurasiya, D. Dong, T.P. Shanley, and Y.Y. Zhao. 2022. Robust genome editing in adult vascular endothelium by nanoparticle delivery of CRISPR-Cas9 plasmid DNA. *Cell Rep.* 38(1): 110196.

Zhu, X.X., J.S. Pan, T. Lin, Y.C. Yang, Q.Y. Huang, S.P. Yang, and D.S. Tang. 2022. Adenine base-editing-mediated exon skipping induces gene knockout in cultured pig cells. *Biotechnol. Lett.* 44(1): 1–18.

2 Current and Evolving Trends in Artificial Insemination of Animal Species

Ranjitha Ashok, Jolitha Adyar Bose, D. Geetha, and Palani Preetham

2.1 INTRODUCTION

Artificial insemination (AI) is one of the blooming sectors where all the research is targeted to bring effective outcomes in animal husbandry, poultry, and agriculture sectors to increase their yield. It is a technique in which sperm is collected, processed, stored, and artificially introduced into the female reproductive system at the right time for fertilization. AI is the most common method for raising intensively farmed domestic cattle, such as dairy cows, pigs, and turkeys. The use of AI in horses, cattle, turkeys, and sheep is growing, and has also been reported in other domesticated species such as dogs, goats, deer, and buffalo. It has also occasionally been used in the conservation breeding of rare or endangered species, such as primates, elephants, and wild cats. AI as a method of reproductive improvement is now accepted and used worldwide. AI is the most important technique ever developed for the genetic improvement of animals. This is possible because a few highly selected males produce enough sperm to impregnate thousands of females per year, whereas embryo transfer can produce only relatively few offspring per selected female each year.

The increasing use of exceptional and proven sires to increase production potential, and to combat natural genital diseases transmitted through services, which contributes to the improvement of animals, is due to the increasing use of AI. AI is a very simple and effective method to increase genetic populations in different species compared to natural seeding and mating technology.

AI is mainly used to get a desired variety of breeds through artificial means, with great genetic stability and efficiency. It helps to understand sperm motility and sperm integrity by the usage of flow cytometer before the insertion into the female reproductive tract. The advancement in research has shown that AI can be used in the field of agriculture to enhance the growth rate of effective crops for seeding. AI is one of the best and easiest ways of achieving the desired variety of breeds, AI prevents the animals from getting infected by infectious disease spread during natural mating under improper conditions. Development of methods for in vitro sperm fertility testing would significantly contribute to conservation efforts.

DOI: 10.1201/9781003452805-2

2.2 HISTORY

The first successful application of AI. He is also known as the "father of modern artificial insemination". In 1922, E.I. Ivanoff, a well-known Russian researcher and pioneer of AI, was the first person to achieve AI in cattle and sheep. In India, AI was first realized by Sampat Kumaran at the "Palace Dairy Farm, Mysore". The first buffalo calf with AI (1943) was born at the Agricultural Institute in Allahabad. Polge, Smith and Parkes discovered the cryoprotective effect of glycerin in frozen sperm technology.

2.2.1 Advantages and Disadvantages of Artificial Insemination

AI in animals was originally developed to control the spread of disease by avoiding the transport of animals with potential pathogens to other animal units for mating and by avoiding physical contact between individuals. The use of semen extenders containing antibiotics also helped to prevent the transmission of bacterial diseases. The advantages and disadvantages of AI are as follows:

AI in the field of biology has played a very important role in the development of various species, increasing the rate of offspring production, and also created a way for the study of new genetics with the species undergoing AI.

AI was introduced mainly to develop the animals disease-free, as the animals were prone to pathogenic diseases due to the transfer of pathogens or viruses during natural mating.

AI has many advantages and disadvantages, which have affected the increase of animal populations with various varieties of the desired genetic traits.

Advantages:

- AI mainly aims to focus on the animal's offspring, as AI provides suitable conditions for the development of healthier offspring.
- AI helps in the control of the spread of pathogens or viruses from one species to another.
- AI prevents the animals from getting infected by the infectious disease spread during the natural mating under unhygienic conditions.
- AI increases the gene population in various species, thus leading to genetic growth and development.
- High yield can be obtained in the animals by the use of prominent genes from males for superior females.
- AI, in terms of environment, can be used in different or specific localities irrespective of weather conditions.
- AI has one special advantage that the semen of the male can be preserved for a long time and can be used in future even after male semen donor's death.
- AI is one of the methods increasingly used in the development of rare breeds and the preservation of endangered species.
- AI in the reproduction process plays an important role in the conservation of sperm by various methods such as cryopreservation, deep freezing, etc., which helps in sperm sexing.

Disadvantages:

- Male shedders can be seen as risk in AI, where these animals shed pathogenic bacteria and viruses in their semen.
- These male animals will not show any clinical traces of contagious diseases during the fertilization process.
- The main disadvantage of these males is that the bacteria or the virus in them will become resistant to the semen, naturally possess antibiotic activity that prevents the formation of biofilms. This causes severe health issues in animals.
- As the AI technique has been increasingly used in animals, the fertility rate in female animals has decreased, especially in cattle and horses, because these animals are more commonly used for AI.
- There may be a loss of many genetic variations in some animals, because selective animals are most commonly used for AI, leading to loss of some original genes.

2.2.2 Prerequisites for AI

AI processes have uniform steps to be followed: selection, collection, preservation, and insertion.

Moreover, several methods are used for the detection of estrus in females, followed by the insertion of semen from a male into the female reproductive tract through a physical method.

2.2.2.1 Collection of Semen

Semen collection is a complex procedure involving coordinated efforts between the animal handler and the collector. Effective harvest of semen involves obtaining the maximum number of sperms of the highest possible quality in each ejaculate. The ultimate objective is to make maximum use of a superior sire, which involves the following:

1. Proper semen collection procedures are used for males.
2. Sexual stimulation to the sire.
3. Proper semen processing techniques.

2.2.2.2 Facilities Needed for Semen Collection

There are essential features that must be considered in designing facilities for collecting semen:

1. Safety of the handler and the collector.
2. he collection area must provide good footing to prevent slipping and injury to the male being collected.
3. Restrain the teaser animals.
4. Easy access for semen collection.

2.2.3 Methods of Semen Collection

1. Collecting the semen from the vagina of the naturally mated female with a spoon, syringe, or long pipette
2. Breeding bag (condom)
3. The massage method
4. Electro-ejaculator
5. Surgical method
6. The artificial vagina (A.V.) method

Disadvantages:

- Small volume of semen, mixed with a large volume of vaginal mucus, is obtained.
- Transmission of venereal disease may occur.
- Injury may occur.
- This method requires a cow in estrus.

2.2.3.1 Breeding Bag (Condom)

It is a strong rubber bag used for semen collection in stallions. It is applied over the erect penis leaving a distance over the glans, and the air should be squeezed out just prior to service. It will be held in place by rubber bands and removed immediately after service.

2.2.3.2 The Massage Method

- It involves massage of the vesicular glands and ampullae by way of the rectum.
- It requires two people, one to do massages and the other to collect the semen; this method is mainly applied for semen collection from bulls.
- After removal of feces from the rectum, a longitudinal back and forth massage is applied. When the urethralis muscle begins to pulsate, the massaging action should be in synchrony with the pulsations.

Advantages:

- No expensive equipment is required.
- This technique avoids the potentially painful aspects of electroejaculation.

Disadvantages:

- Skilled experience is required.
- Libido, mating ability, penile erectile function, and the ability to ejaculate are not evaluated. Because erection does not occur, semen samples may be contaminated with epithelial cells, bacteria, and dirt, especially when it dribbles through the prepuce.
- Urination may occur during the massage.
- Semen volume and concentration are less than the true ejaculate because it also includes the residue of previous ejaculates accumulating in the ampulla.

- Urine residue is also present, because of the absence of the bulbourethral gland's secretions.
- Rectum injury may occur.

2.2.3.3 Electro-ejaculator

It is used with impaired bulls that are unable to mount, that are not accustomed to being handled, and that have low sexual activity, and also for aged males. Electro-ejaculator is designed to stimulate the pelvic sympathetic and parasympathetic nerves (Figure 2.1).

2.2.3.4 Electro-ejaculator Procedure

- The rectum is emptied of feces; longitudinal massage is applied over the ampulla and urethralis muscle for 1–2 min.
- Lubricated probe introduced into the rectum with the electrodes facing ventrally.

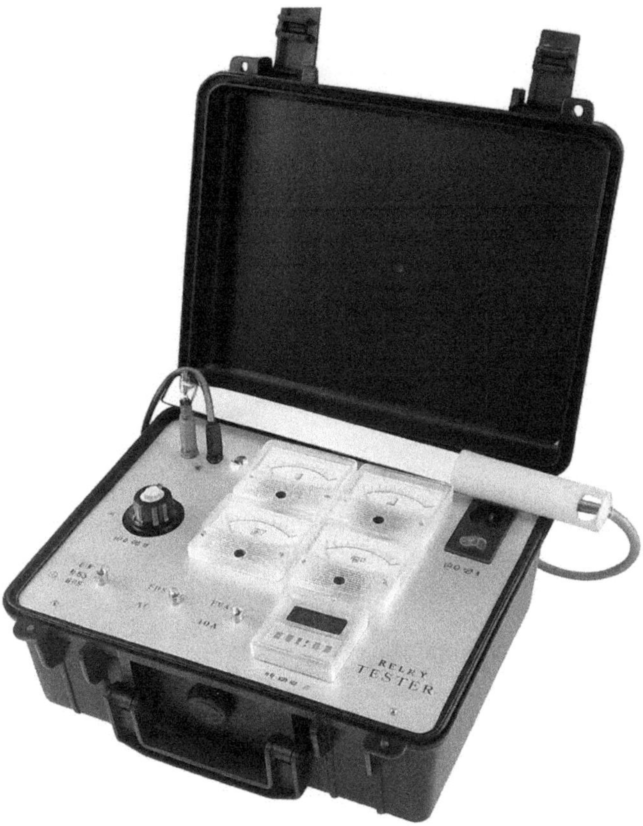

FIGURE 2.1 A visual representation of an electro-ejaculator used for impaired bulls.

- Electrical stimulation should begin slowly until the bull shows a minimal response, then gradually increase in intensity. The stimuli should last 1–2 s and then be discontinued for 0.5–1 s.
- After several stimulations, clear pre-seminal fluid begins to flow from the protruded penis.
- As soon as the cloudy sperm-rich fraction begins to flow from the penis, the sample is collected in a collection cone with the test tube placed over the penis.
- Rectal probe removed.

Advantages:

- Can be used on a wide range of bulls and rams.
- Can be used on untrained bulls and rams.
- Usually good collection volumes.
- Can be used on bulls unable to mount.
- Can be used for initial inspection of bull's penis and sheath, to check for any growths or extension problems.
- Can be used on bulls with decreased libido.
- Electroejaculation is the only technique useful for collecting semen from wild animals, in which the male is anaesthetized prior to the procedure.

Disadvantages:

- Increasing voltage too rapidly can result in ejaculation without an erection, so the semen will be contaminated with preputial dirt.
- The nerves of the hind limbs are also stimulated by electrical current, which results in stiffening of the legs and the animal may go down.
- Collected semen quality is large in volume but low in concentration, and sometimes contaminated with urine.
- This method is inhumane because it causes severe pain to the animal.
- Electrical stroking of the animal may occur as well as the operator.

2.2.3.5 Surgical Method

A small incision is made in the perineal region to locate the urethra, and then a cannula is inserted into the urethra. This method is impractical because it leads to adhesion and constriction of the urethra as well as infection of the urogenital system.

2.2.3.6 The Artificial Vagina

- The A.V. method of collecting semen is the fastest and more sanitary of the various methods available.
- An A.V. is used to collect semen from many species of animals, most prominently cattle and horses, and, also from sheep, goats, rabbits, and even cats.
- An A.V. provides a good imitation of the natural vagina, which uses thermal and mechanical stimulation to stimulate ejaculation.

2.2.4 Components of A.V.

1. A strong outer casting (cylinder)
2. Thin inner rubber (sleeves)
3. Rubber bands
4. Valve
5. Receiving cone
6. Test tube
7. Warming bag (Figure 2.2)

Advantages

- Practically, the whole ejaculate is collected in the uncontaminated and natural stage.
- It is free from the extraneous secretions.
- The sterile condition of the apparatus ensures disease control.
- The viability of the sperm is better.
- No female is needed if a dummy is available.

Disadvantages

- It takes a lot of time, especially in breeds with decreased libido.
- Bulls need to be trained.

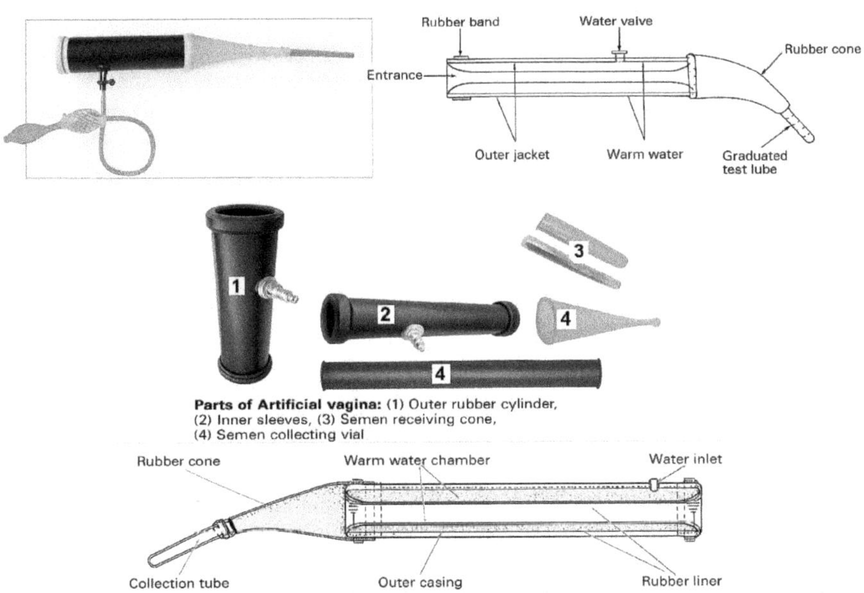

FIGURE 2.2 A pictorial representation for understanding the components of the artificial vagina: (1) outer rubber cylinder, (2) inner sleeves, (3) semen-receiving cone, and (4) semen-collecting vial.

- Increased labor.
- Cannot be applied on structurally unsound bulls.
- Can be hazardous to the collector.

2.2.5 Constituents of Semen

Semen is the main component to be considered in the process of AI. It consists of millions of sperms termed spermatozoa, and seminal plasma, a watery fluid. Seminal plasma consists of different accessory glands like the seminal vesicles, prostate gland, and bulbourethral gland. These gland activities vary from species to species; for example, in primates, after the collection of semen through ejaculation, some semen gets coagulated very quickly and then gets liquefied within half an hour.

But in some species, the semen will not get liquefied in vitro. To avoid this, the enzymes are added, such as hyaluronidase, fibrinolysis, trypsin, and collagenase, which cause acrosomal damage in spermatozoa that will be used for AI. Camelid semen has increased in volume by a 1:1 ratio by liquefying in 60–90 min at 37°C, according to a recent study. Seminal plasma is the one which contains all the beneficial requirements to complete the fertilization. It contains fructose as an energy substrate to support stability, along with proteins and vital ions such as bicarbonates, calcium, magnesium, and zinc—all crucial for facilitating sperm transport.

Seminal plasma will also be maintained in a lethargic state in the epididymis, but it will be used to transport spermatozoa into the reproductive tract of females by allowing the sperm to swim to the locus of fertilization.

Sperm adhesins and CRISP (cysteine-rich secretory proteins) are the proteins present in seminal plasma which are important for maintaining sperm fertility. These proteins usually bind to the spermatozoa immediately after the release of sperm from male species in a sequence of intracellular incidents through a second messenger pathway.

In some species, prostasomes, vesiculosomes, and epididymis have been identified in seminal plasma from the accessory gland based on their origin, and fuse with the outer membrane of sperm, which increase the sperm's fertilizing ability by increasing motility and are involved in increasing sperm capacity.

Seminal factors promote sperm survival in the female reproductive system, modulate the woman's immune response, tolerate fertilization, and improve the condition of the uterine environment for embryonic development and the endometrium for implantation. The mechanism of action in the endometrium involves the recruitment and activation of macrophages and granulocytes as well as the remodeling of dendrites, thus improving the sensitivity of the endometrium to the implanted embryo. The release of cytokines has embryotropic properties and can also affect tissues outside the reproductive system. Contact with spermatozoa triggers cytokine activation within the uterine fluid and the epithelial glycocalyx that lines the uterine lumen. These cytokines interact with the developing embryo as it passes through the fallopian tubes into the uterus before implantation (1992). Human clinical trials have shown acute and cumulative benefits from sperm exposure, but also through a partner-specific pathway. Birth rates among couples undergoing infertility treatment improve when women have sexual intercourse shortly before embryo transfer.

The use of seminal plasma pessaries has been reported to improve pregnancy success in women with recurrent miscarriages (Coulam and Stern, 1993). The specificity of the partner response suggests an increased incidence of preeclampsia in pregnancies resulting from donated eggs or sperm without prior exposure to donor sperm or fetal antigens (Table 2.1).

2.2.5.1 Semen Processing

Seminal plasma plays an important role in activating spermatozoa and in the female reproductive tract, where it provides the stability to sperm for long-term survival outside the body. Under physiological conditions at ejaculation, seminal plasma activates spermatozoa, prompting them to migrate away from the site of semen deposition within the female reproductive tract. Hence, it is a prevailing practice to add a semen extender to the semen, to dilute toxic elements in seminal plasma, which provide nutrients for the spermatozoa during in vitro preservation and to buffer their metabolic by-products. Semen can be given in doses by dividing it into several semen doses by adding the extender.

These semen doses in specific conditions consist of a number of spermatozoa that have been ascertained to be good for fertility in artificially inseminated females. This helps to give the proper dosage to each individual based on their capacity in required proportions, enabling them to achieve the best fertilization.

2.2.5.2 Semen Preservation

Semen preservation can be done by various methods for the future use. It can be stored or used fresh. Semen, which is used immediately after collection, is called fresh semen, for example, turkeys, human beings; semen, which is stored at a reduced temperature or by artificial means, is called stored semen, for example, horses, pigs, dogs; or preservation of semen after freezing and thawing is called cryopreservation, for example, bulls.

These are the processes used to store the semen to get a new variety of species with desired genetic variation.

2.2.5.3 Fresh Semen

In comparison to animal species, human semen cannot be extended prior to processing, and it cannot usually be kept for more than a few hours before the usage. Human semen gets coagulated at the earliest, so the semen is used fresh for good fertilization. Poultry semen cannot be extended as much as it is commonly done for other species. Spermatozoa are prone to early dilution by the increased adverse effects. Special cases include some animals like goats that have very sensitive semen which cannot be stored for long time especially at 37°C, as it contains an enzymatic component of the bulbourethral gland secretion that hydrolyzes milk triglycerides into free fatty acids, which will directly affect the sperm motility and the integrity of outer membrane of bulk spermatozoa. Goat semen can be stored in liquid medium for example, liquid nitrogen at 4°C, but still the semen fertility is retained for only 12–24 h. The rate of extension, which is used for stallion semen, varies between countries but rates of 1:2, 1:3, or even 1:4 (v/v) semen:extender is very common. The standard practice in some countries is to have 500 million or 1 billion progressively with respect to motile

TABLE 2.1
Semen Composition of Different Domestic Animals

Semen	Bull	Ram	Goat	Boar	Stallion
• Dry weight (%)	9.5	14.8	–	4.6	4.3
• pH	6.48–6.99	5.9–7.3	–	6.85–7.9	6.2–7.8
• Specific gravity	1.035	–	–	–	1.013
• Sperm Concentration (10^6 /ml)	300–2000	2000–5000	1000–5000	25–350	30–800
• Sperm atocrit(%)	10	33	–	2	3
• Volume (ml)	2–10	0.5–2	0.5–2.5	150–500	20–300
Seminal Plasma					
• Acetyl glucosaminidase (units/ml)	15000	16000	–	–	–
• Ascorbic acid (mg/100ml)	8.7	5	–	–	–
• Bicarbonate (mmol/l)	7.1*	7.1*	–	–	–
• Calcium (mmol/l)	9.3	1.9	–	–	6.5
• Chloride (mmol/l)	49	18	–	96	–
• Citric acid (mg/100ml)	357–1000	137	–	36–325	8–53
• Ergothioneine(mg/100ml)	Trace	Trace	Absent	6–30	3.5–13.7
• Fructose (mg/100ml)	120–540	150–600	–	20–40	<1
• Glutamic acid(mg/100ml)	35–41	76	–	–	–
• Glycerylphosphorylcholine(mg/100ml)	110–500	1600–2000	1400–1600*	110–240	40–110*
• Inositol (mg/100ml)	25–46	10–15	–	380–610	19–47
• Magnesium (mmol/l)	3.4	2.4	–	–	3.8
• Mannosidase (units/ml)	400	50	–	–	–
• Potassium (mmol/l)	44	23	–	16	26
• Protein (g/100ml)	3–8	–	–	–	–
• Sodium (mmol/l)	117	78	–	122	114
• Sorbitol (mg/100ml)	10–136	26–120	–	6–18	20–60

stallion spermatozoa for fresh or cooled semen doses, respectively. Boar semen doses contain more than 3 billion progressively motile spermatozoa.

2.2.5.4 Stored Semen

Storing semen is one of the techniques to preserve the required or rare semen for future use. Stored semen can extend semen at reduced temperature, which helps to extend sperm life by reducing their metabolism as well as by inhibiting bacterial growth and reducing pathogenic effects. Bacteria, which grow in the collected semen by utilizing the nutrients in semen extenders, mainly used to increase the lifespan of the semen for about a year or more, start competing with spermatozoa for these limited resources. Given the limited resources, it effectively sustains a suitable environment for semen preservation. The stored semen releases metabolic by-products, which help maintain conditions that minimize contact-based degradation, thereby safeguarding the viability of spermatozoa.. Prolonged storage of semen leads to a progressive alteration in the resident bacterial community. Besides, as bacteria die, they may start to release endotoxins into the collected semen that can be toxic to spermatozoa. However, cooled stored semen is one of the methods of choice for breeding horses and pigs, which enables the semen doses to be transported to different geographic locations for AI. Stallion semen is stored at approximately 6°C while boar semen is stored between 16°C and 18°C. This semen will be preserved under cold conditions by deep freezing, which allows the collected semen to stay viable for longer durations.

Most of the boar semen doses are sold as cooled doses to avoid contamination and to prevent contagious disease. In contradictory to this, some stallions can produce spermatozoa that cannot tolerate cooling, especially deep freezing. They began to rapidly lose progressive sperm motility. In such cases, the only option currently used is to use fresh semen doses for AI immediately after semen is collected from the male donor, whilst a new method of processing, called centrifugation by a single layer of colloid, has been shown to solve the problem, which will be discussed later. Therefore, this stored semen plays an important role in future applications to enhance the quality of breeds.

2.2.5.5 Cryopreservation

Semen, which is collected by the male donor, is the most useful element for AI, and it can be cryopreserved, after the collection of semen through artificial methods. This method of preservation flawlessly enables the collected semen to be stored for an unlimited period of time without the loss of the sperm quality until needed for AI. Since the frozen semen does not degenerate, it can be quarantined until the male donor has been shown to be free from contagious disease at the time of semen collection. However, the spermatozoa from collected semen of various species vary in their ability to withstand the temperature under cryopreservation: ruminant spermatozoa will survive well, whereas the poultry spermatozoa do not survive for a longer period, with less than 2% maintaining their fertilizing ability on thawing. This is the reason why poultry semen collected is highly difficult to store for a longer period (Wishart, 1985). In animal husbandry, the cost of cryopreservation and the likelihood of AI success should be considered when deciding whether to use doses of fresh, chilled, or frozen semen, because the sperm rate and motility will decline very soon. Sperm are mixed with a protective solution, containing lipoproteins, sugars, and a cryoprotectant such as glycerin. These ingredients help maintain the integrity of the membrane during the cooling and heating process, which enables to protect the integrity of sperm. However, the motility of the sperm must also be maintained so that the sperm thawed after insemination can reach the eggs and fertilize them at the earliest without causing any infection. In most species, seminal plasma is removed by centrifugation to avoid the unwanted debris before mixing with a cryosolvent, for example, stallion, boar, goat, and human sperm. The stretched sperm is packaged in straws and frozen in liquid nitrogen vapor and then submerged in liquid nitrogen for long-term storage which allows the semen to survive for longer durations until further use for AI in desired animals for better breeds. Despite extensive research into the components of cryoextension and cooling and warming rates, there is a significant variation in the success of sperm cryopreservation between various species. Human sperm can be frozen relatively efficiently using commercially available cryo-expanders and programmable freezers because human sperms are very delicate.

2.2.6 Estrus Detection and Ovulation

AI is one of the methods involved in the fertilization process by using high-grade semen from a male to a superior female. At the time of fertilization, it is mandatory to know the exact ovulation period of the desired female to get the best breed of their choice. The effectiveness of AI also depends on the deposition of sperm in the woman's reproductive tract at the time of ovulation because deposition of sperm plays a key role in fertilization at the right time. Like humans, some domesticated animals breed all year round, such as cattle and pigs, but others have a specific period of reproductive activity called the breeding season because there is a variety of species which differ each other in their reproduction process, such as sheep and horses. The start of the breeding season is controlled by the photoperiod. Both patterns of reproductive behavior are characterized by waves of ovarian activity that peak at ovulation. However, in some other species such as cats, rabbits, and camels, ovulation occurs in response to the mating stimulus. In species that have spontaneous ovulation, ovulation occurs during or shortly after estrus, the time when the female is receptive to the male. Since the success of AI depends on sperm being deposited at the right time relative to ovulation, heat detection is critical if the female is to be fertilized at the right time. Conspecifics are of course very good at recognizing females in heat, but since in many insemination farms, there are no males around, it is important that breeding staff can recognize heat behavior; if the heat detection is missed, the entire AI process will go wrong by wasting both time and money.

Although some animals may exhibit well-developed estrous behavior, for example dairy cows, others will not. Behavioral signs of estrus in cows include restlessness or increased activity, vocalization, resting chin, vulvar swelling, vaginal discharge, and climbing on other cows. However, there are differences in the frequency and intensity of these symptoms between races based on their reproduction time. Sheep and goats may experience vulvar swelling and vaginal discharge and are usually male-biased. When using AI in sheep, estrus is usually synchronized with hormones: progestogen-soaked

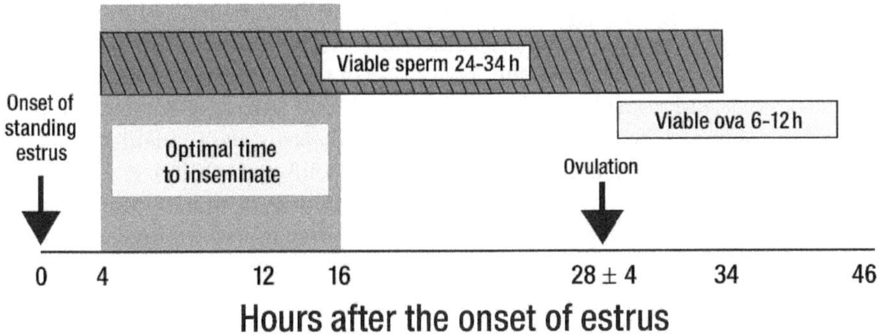

FIGURE 2.3 Biological events prior to deposition of semen in the selected female: (1) sperm is viable for 23–34 hours, (2) at 0 hour, the onset of standing estrus begins, (3) optimal time to inseminate/deposit the sperm is between 4 and 12 hours since the onset of estrus, (4) ovulation occurs at 28 hours (±4), and (5) ovum is viable for 6–12 hours.

vaginal sponges are inserted into the vaginal area of the sheep to suppress the sheep's natural ovarian cycle for about 11–12 days. When the sponge is removed, gonadotropin from the pregnant mare's serum is administered and then AI is performed at a specific time. Alternatively, a vasectomized ram carrying the marker can be mated with females. When the females are in heat, the vasectomized ram marks them during mating, allowing them to be identified by AI for the better fertilization to occur in them. Sows and mares in heat can be recognized by their behavior toward irritating males. These are some of the methods followed to know the ovulation period in females for the effective fertilization to happen during the AI process (Figure 2.3).

2.2.6.1 Induced Ovulation

Ovulation is the main process required for the AI process to happen where it varies from species to species. It is mandatory to know the ovulation period in females for the ovule induction process.

When performing AI in species that normally undergo ovulation induction, such as rabbits, cats, and camels, ovulation stimulation is required for the species during their ovulation period. The easiest way to achieve this stimulation is to mate a female with a vasectomized male. However, this practice is undesirable from a disease control perspective, and it requires vasectomized males. The most acceptable alternative is the administration of luteinizing hormone, usually in the form of human chorionic gonadotropin. The main disadvantage, however, is that repeated injections of this foreign protein can cause the female to develop antibodies, thus inactivating subsequent doses. These external insertions to the reproductive tract of females will even cause infections sometimes in the female reproductive part.

2.2.6.2 Artificially Induced Ovulation

Induced ovulation should be done to get the correct ovulation time of the female at the time of the AI process because it is a crucial step for female for the proper fertilization. To ensure this, the spontaneous ovulators can receive hormones to ensure that ovulation occurs at the right time with regard to AI. However, the use of hormones in animals intended for animal feed has been banned in the European Union since 2006, and local regulations may also apply in other parts of the world, because there was a loss of yield and a decrease in the gene population. Until now, most dairy goats in France have been inseminated with frozen semen outside of the breeding season, after hormone therapy has induced estrus and ovulation. This protocol provided a hit rate of approximately 65% Leboeuf et al., 2008). An alternative to administering artificial hormones is off-season breeding, which involves changing the photoperiod or adding a goat to the herd. The practice is also widespread in intensive sheep farming, as it is a natural practice with a low rate of spreading infections and disease in animals, and also gives healthy offspring.

2.2.6.3 Deposition of semen in the female

The collection of the semen deposit is important in a controlled environment. Deposition of the semen without pathogens is very crucial to prevent the spread of disease. There are species differences in the site of sperm deposition during natural mating. In ruminants and primates, sperm is deposited in the vagina or cervix, whereas in pigs, dogs, camels, and horses, sperm deposition occurs intrauterine. In most types, an insemination catheter can be inserted through the cervix to allow sperm to attach to the uterus for fertilization to occur during AI. An exception are sheep and goats, in which the tightly folded cervix does not allow the easy passage of the insemination catheter which making it difficult to use them for the AI process. The advantage of sperm storage in the uterus is that the sperm have to travel a shorter distance to reach the fallopian tubes, and fewer sperms sent there will be lost due to retrograde flow. In the AI process, the stored semen are used in the form of insemination dose which contains smaller amount of sperm than vaginal deposition, allowing the ejaculate to be divided into multiple doses of different concentrations. In the AI process, the semen has to pass by the cervix, which can be a barrier to sperm flow. The disadvantage, especially with intrauterine inseminization (IUI) in humans is that the seminal plasma is also introduced into the uterus unless special measures are taken to separate the sperm from the seminal plasma before the IUI, because seminal plasma provides the ions and proteins contained in it which are involved in better fertilization.

2.2.7 Artificial Insemination (AI) in Cattle

AI is the process of collecting sperm cells from a high-merit male animal and manually depositing them into the reproductive tract of a female. Once the semen is collected by artificial methods, it can be used fresh or stored by various methods. There are several potential benefits from the use of AI, including using it for the conservation of rare breeds and endangered species.

2.2.7.1 Increased Efficiency of Bull Usage

AI has become more beneficial in the field of dairy industry; therefore, bull usage in AI has become the common step to get a good breed in terms of yield and efficiency. During natural reproduction, the male deposits more sperm than would theoretically be necessary for pregnancy. In addition, natural farming is physically demanding.

These two factors limit the number of natural matings a male can perform. However, the collected semen can be diluted and expanded to produce hundreds of doses from a single ejaculate, as it contains enough amount of semen containing all the required proteins and ions for fertilization. In addition, sperm can be easily transported; This allows for the simultaneous fertilization of multiple females in different geographical locations and the storage of sperm for extended periods of time, meaning that males can continue to produce offspring long after their natural reproductive life has ended. Because bull semen can be stored for years together without any physical or physiological changes, it maintains it sperm integrity even after long-term preservation.

2.2.7.2 Increased Potential for Genetic Selection

The increased potential for genetic selection leads to the decrease of male animals, especially those that are of low-grade merit, and this leads to the decline of other species or causes species to become endangered. Because males can produce more offspring through AI, fewer males are needed. Therefore, only some of the best males can be selected as parents, increasing the intensity of selection. Additionally, because males can have multiple offspring, their offspring can be used in an offspring testing program to more accurately assess the male's genetic merit. Finally, through AI, individual breeders can increase the gene pool into which their animals can be bred, potentially reducing the effects of inbreeding. This is one of the disadvantages in the field of AI, especially in animals.

2.2.7.3 Decreased Costs

Maintenance of male animals is cost-consuming compared to females but if the male is of high-grade merit, it would be beneficial, or else, it would consume lots of resource for the development. Males are often larger than females and can consume relatively larger quantities of food. Additionally, males are often stronger and more powerful and may misbehave, requiring special housing and handling equipment compared to females.

2.2.7.4 Increased Safety for Animals and Farmers

Male animals can grow large and aggressive. Because of these factors, keeping a bull on a farm can be dangerous. Additionally, due to the relatively larger size of adult males compared to females, natural mating is more likely to cause accidents and injuries to the cow or bull than AI. Animals should be kept under good housing facility where it should not harm other animals either physically or resourcefully. Animals which are more aggressive should be always kept away from the farmers, as they may harm farmers, and sometimes may cause death.

2.2.7.5 Reduced Disease Transmission

Sexually transmitted diseases can be transmitted between males and females through natural mating. Some pathogens can be transmitted through sperm during AI, but the collection process allows for analysis of the pathogens. The quality of the sperm obtained is also checked regularly, which can help avoid problems associated with male infertility. However, there are some potential disadvantages of AI that should be taken into account. First, it may require more work. Men instinctively recognize women who are in good condition to conceive. In the case of AI, identification is the responsibility of the breeder. Poor detection leads to reduced fertility rates. Furthermore, increasing the number of offspring per male only has selective advantages if the best males can be accurately identified. Otherwise, this process only reduces the genetic variability of the population. Increasing the number of children per man always reduces the gene pool. The collection of semen deposits from male must be used fresh with certain animals to avoid infections. The stored semen should be preserved at controlled environment by using cryopreservants to avoid the risk of pathogens and bacteria to reduce the disease in animals used in AI. The stored semen is separated according to the proper dosages; only required dose should be given to animals. The benefits of more intensive selection must be weighed against the negative effects of reduced variability.

2.2.8 Artificial Insemination (AI) in Pigs

Artificial insemination (AI) has a wide range of applications in hybrid breeding technologies across various species. Among animals, it is most prominently used in pigs because of its genetic variation in gene population.

Pigs are used extensively in the biotechnology research purposes, as they have shown the best results in all the fields of biological research, for example, transplantation, xenobiotics, and regenerative drugs.

AI of pigs is widespread in countries with intensive pig production. In Western Europe, more than 90% of sows have been bred using AI for more than 20 years. Compared to natural selection, AI is a very useful tool for introducing better genes into sow farms with minimal risk of disease (Maes et al., 2011). The result of AI depends largely on the quality of the sperm and the fertilization process. In practice, fresh, diluted semen is often used for intracervical insemination in pigs. The semen is obtained from boars present either in farms or in centers specializing in in vitro fertilization (IVF). These provide a variety of breeds and genetic lines, distributed as ready-to-use semen doses of consistent quality to various sow farms.

There are three important aspects to consider. First, only semen from healthy boars should be used, as diseased boars may expel semen contaminated with pathogens. The semen from commercial AI centers is sent to numerous sow farms. Contaminated semen can therefore lead to rapid transmission of pathogens and epidemics in many different sow farms. Therefore, pig AI centers have strict regulations and guidelines to prevent the spread of diseases. The second important aspect is the fertilizing power of the sperm doses produced. The fertilization potential of a sperm dose is inextricably linked to the quality of the sperm itself. It is therefore necessary to test ejaculations. The third important aspect of AI centers is the sperm processing procedure. This is important not only because of the low occurrence of microorganisms, but above all for obtaining high-quality sperm, that is, live sperm in ready-to-use sperm doses that can be used for several days. The dilution procedure and handling of the sperm, the properties of the diluent, and the sperm microenvironment affect sperm survival and longevity.

Pig AI is widely used and is a very useful tool for introducing high-quality genes into sow farms with minimal risk of disease transmission. In practice, fresh, diluted sperm (3 billion sperm in 80–100 mL) is often used for intracervical insemination.

The success of AI is largely determined by the quality of the sperm and the fertilization process. Various parameters and techniques can be used to assess sperm quality. Although more advanced technologies provide more accurate information, sperm quality in AI centers is assessed primarily on the basis of concentration, morphology, and motility, using techniques that are simple, inexpensive, and practically easy to implement. Critical issues in pig AI include detection of sow estrus, timing of insemination, and application of strict hygiene measures. Future developments will focus on new technologies to better assess sperm quality in practice, preserve them for longer periods, and efficiently inseminate sows with fewer sperm using new AI techniques.

The pig semen after the deposition will be stored by cooling, deep freezing, and cryopreservation methods to increase the shelf life of sperm for long-term use. This stored semen would be used in doses to get a better breed with increased gene population.

2.2.9 Artificial Insemination (AI) in Poultry

Another field where AI is used is poultry. As the demand for poultry, primarily chicken and turkeys, is increasing worldwide, farmers continue to expand poultry farming, to increase the yield in terms of both number and efficiency.

AI is often used on commercial turkeys. This is primarily due to the selective breeding of commercial turkeys with larger body weights and broader chests, as well as the inability of males to continuously transfer sperm to the hen during mating. The broiler industry has not brought AI to the level of the turkey industry, but it is sometimes used in breeding lines and in areas where labor is relatively cheap. To measure the outcome of of AI in breeding turkeys, a hypothetical flock of 500 breeding hens were inseminated with 100 μl of diluted semen (1:1) twice weekly before egg production, then once weekly for 24 years hours weekly egg production would require 13,000 fertilizations with 650 ml of sperm. It is clear that collecting semen and inseminating hens is laborious, because both male and female require weekly care.

Looking at the history of AI use in turkey breeding, it is safe to say that in the 1960s, weekly inseminations were based on the amount of semen per dose of undiluted semen. In the 1970s and early 1980s, farms began diluting sperm and fertilizing a certain number of sperms per dose. In the mid-1980s and 1990s, hens were initially inseminated a week before laying began, with a known number of "live" sperms. Even today, fertilization before, egg production is widespread; Most, but not all, companies returned to insemination at semen volume or sperm count per dose that were known in the 1970s and 1980s. Sir Peter Lake has written an excellent overview of the history of AI and its impact on the poultry industry and what is needed to accelerate the application of AI in poultry. Unfortunately, AI technology has not made much progress since this review article. More than 15 years later, the only significant progress concerns the assessment of sperm motility and implications for paternity in men who produce highly motile sperm. Regardless, research into the biology of poultry sperm and the cellular and molecular basis of transport, selection, and storage of oviduct sperm can be expected to lead to the following innovations in AI technology at some point in the future: expansion of intervals up to 10–14 days between inseminations (compared to 7 days) with fewer sperm per insemination, preservation of sperm in vitro for 24–36 h at room temperature with minimal loss of sperm viability, and the possibility of obtaining transgenic offspring after fertilization with sperm-carrying transgenes.

Recent studies have shown that even broiler industry has started to use AI and achieved effective results such as increase in the weight of chicks and their number.

2.3 ARTIFICIAL INSEMINATION (AI) IN HORSES

AI in horse s has become a popular breeding technique, as it allows horse breeders to improve the genetic quality of their herd without the risks of natural breeding. AI in horses involves the collection, preservation, and transfer of semen from a stallion to a mare. AI is helpful when a particular stallion may live far away from the mare, or when a mare has been difficult to impregnate. The AI process allows for more control in the breeding of a mare than natural mating or, pasture, breeding.

The success rates of AI in horses can vary depending on several factors, such as the quality of the semen, the timing of insemination, and the reproductive health of the mare. On average, success rates for AI in horses are between 50% and 70%. Frozen sperm generally has a lower success rate than refrigerated sperm due to the loss of integrity and motility rate of sperm by the usage of more chemical cryopreservants.

Possible reasons for lower success of AI include the following:

- Stored semen can cause poor motility or morphology of sperm.
- Stored semen can decline the integrity of sperm.
- Due to loss of ions, there is delayed transport of sperm.
- Due to lack of reproductive organs hygiene there is a poor uterine health (thickening or other concerns, often diagnosed during a breeding soundness exam upon biopsy and culture).
- Carelessness of breeders to know their heating time.
- Improper timing of AI relative to their ovulation.
- Improper induced ovulation.
- Reproductive status or history of a mare.
- When was the last time she foaled?
- Is she a young maiden mare, or has she had multiple foals?
- Older mares who have never had a foal before often have a harder time conceiving and maintaining a full-length pregnancy.

The best conditions for AI success include the following:

- Healthy uterus and sperm with good motility (pre- and post-shipping) and morphology in male donor
- Exact ovulation time
- Usage of proper dose of semen
- AI done in correct fertility window for the type of semen being used
- Hormone assistance for ovulation timing and/or uterine clearance after breeding
- Pre- or post-uterine hygiene as needed
- Luck!

It is important for both handlers and veterinarians to remember that mares are not machines. Even if all requirements are correctly met, there is no guarantee of a successful pregnancy. If you decide to breed mares, it is important to prepare mentally and financially for the possibility that something won't work after the first, second, or even third round of attempts. Sometimes things are beyond the vet's control and changes may be necessary, including the type of semen used, the stallion used, or repeating the mare's uterine diagnostics to ensure she is a good candidate for breeding.

AI offers many potential benefits for horses, particularly when it comes to mare safety and transportation. Intact stallions bred live can be potentially dangerous for everyone involved, both the horses and their handlers. If the mare is not yet fully ready to cover, she may push away from the stallion, which may involve screaming and kicking; and stallions often forget their manners in the heat of battle. Helmets are important for riders because the stallion often kicks when attempting to mount the mare and both horses may decide to kick. Breeding can also sometimes cause injury to the mares' reproductive system, and stallions often bite or inflict wounds on the mares' flanks.

Although the success rate with AI may be slightly lower than with live mating, safety considerations are important, and it is much easier to ship semen overnight than to send a mare to a possibly different state where she will remain there during the breeding season until a success is achieved.

One of the biggest disadvantages of AI compared to mating a live mare with a nearby stallion is the cost. The fees for breeding artificial animals include the breeding contract, veterinary examinations, sperm shipping fees, and any unexpected complications or repeat fees if everything does not go as expected. The most common obstacles include the problems discussed above, as well as unexpected delivery delays or problems, missed ovulation, and failure to submit the "list" on time on the day of collection (each semen collection includes only a certain number of doses, which are distributed on a first-come, first-serve basis).

2.3.1 AI in Other Species

The AI of alien species presents some new challenges compared to native species. In many cases, little is known about a species' reproductive biology and handling animals can cause stress and pose a risk of injury. Animals must be managed appropriately to achieve and maintain pregnancy. Therefore, the animals must be handled with suitable care in terms of handling, feeding, and also the housing. Successful applications of AI on deer, buffalo, and camelids have been reported.

2.3.2 Future Trends in AI

The use of AI in agriculture is expected to continue to increase to get a better yield in crop varieties. In addition to enabling more effective and efficient animal production, AI can also be combined with other emerging biotechnologies such as cryopreservation, solid sperm selection by monolayer centrifugation, and sperm sex selection. These are the various methods that have been advanced in recent days for the storage of semen with good quality.

2.3.3 AI in Increasing the Efficiency of Livestock Production

Apart from some specialized sheep and goat farms that produce milk for cheese and orient toward intensive meat production, breeding of these species is generally limited to marginal areas not suitable for agricultural production or livestock grazing. The choice of production features was limited. However, they are now attracting growing interest in developed countries, as there is growing awareness that small ruminants can make better use of limited resources than large ruminants such as cattle and produce less methane and wastewater. In many developing countries, sheep and goats are better adapted to the climate than cattle, and it is also culturally acceptable to eat their meat and dairy products. It is therefore likely that we will see a strong use of AI in sheep and goats in the future, with particular attention to improving production characteristics through the introduction of better genes. However, it is important that any AI program aimed at large-scale improvement of national heritage is supported by improvements in animal husbandry and health, otherwise AI-induced pregnancies will not be carried to term and the offspring will not survive or develop. Many advanced ART methods are of little use in areas where basic screening knowledge is inadequate. The effective knowledge only can help to grab the opportunity of AI to get the desired variety of breed with the increase in the level of genetic stability and effective growth.

2.3.3.1 Biomimetic Sperm Selection

A potential drawback of AI is that, if used, it can bypass the natural selection mechanisms of the female reproductive system, which aim to select the best sperm for fertilization. Biomimicry is the use of technologies and/or processes that mimic a natural event. Several in vitro methods have been proposed that could mimic the selection of high-quality sperm in the female reproductive system and would therefore meet the definition of biomimicry in ART. These include sperm processing methods such as swimming, sperm migration, filtration, and colloid centrifugation (reviewed in Morrell and Rodriguez-Martinez, 2010). Among these methods, colloidal centrifugation is the most suitable method for livestock and human sperm selection. The biomimicking is the best method used to provide the required environment for the desired breed during the fertilization process.

2.3.3.2 Density Gradient Centrifugation

The stored semen or the fresh semen which is deposited must be centrifuged before the use to avoid the transformation of debris into the female reproductive tract. This helps in the removal of chemical agents and also the spread of contagious disease from the male donor. Human sperm intended for the treatment of infertility is generally processed to remove seminal plasma and select the best-quality sperm. In most cases, this is achieved through sperm migration, which separates the most motile sperm from the rest of the ejaculate, or through density gradient centrifugation (DGC), which selects the strongest sperm. The advantages of DGC include the following (Morrell, 2006):

Slow-moving and abnormal sperm are eliminated.

ROS sources (cellular debris, leukocytes, epithelial cells, and dead or dying sperm) are removed.

Sperm survival improves with frozen storage compared to nonfrozen storage.
Bacterial contamination is combated without the use of antibiotics.
The more sperm integrity, the more will be the healthy offspring.

2.3.3.3 Single-Layer Centrifugation

DGC is rarely used in animal sperm processing due to the limited amount of sperm that can be processed at once and the time required to prepare different layers. A new sperm preparation technique, single-layer centrifugation (SLC) by colloid, has been developed at the Swedish University of Agricultural Sciences (SLU) to select the strongest sperm from the ejaculate. This method is similar to DGC, but is more suitable for animal sperm as it has been extended to processing whole ejaculates. The main applications of SLC selection are similar to those of DGC and have been extensively studied by Morrell and Rodriguez-Martinez (2010). The SLC helps to have the strongest sperm to swim into the female reproductive tract to reach faster for the fertilization to get a healthier breed.

2.3.3.4 Sex Selection

Sex selection is the important step in the AI process. One wrong selection may account for a number of unhealthy or genetic disordered progeny which may continue for generations.

For many centuries, breeders and researchers have tried to control the sex of their offspring for various reasons. Originally, male offspring were preferred for meat production due to males' better feed efficiency and lean fat percentage, while females were preferred for milk purposes, except that some males with high genetic value were still necessary as bulls. Couples may want their child to be a specific gender to avoid gender-related disorders.

Many methods have been proposed to separate spermatozoa with X and Y chromosomes based on their physical properties, for example the size of the sperm head, or functional properties, for example swimming speed. However, the only method that has proven effective is the selection and separation of sperm cells whose DNA is stained with the bisbenzimidazole dye H33342 using the sorting capability of a flow cytometer which enables to identify the best sperm motility ratios (Morrell et al., 1988; Johnson et al., 1989). This method works because X chromosome is larger than Y chromosome, so it absorbs more DNA-specific dye and shows higher fluorescence when sperm pass through a laser beam. In bulls, for example, the difference in DNA content between the X and Y chromosomes is approximately 4.2%. However, sorting sufficient quantities for a fertilization dose takes too long in a flow cytometer because the stained sperm must pass through the laser beam one at a time to detect the DNA content. In addition, the pregnancy rate after insemination with sexed bull semen is lower than with unsexed semen, making the procedure ineffective and expensive. It has been shown that staining profiles are very individual, and it is therefore not possible to effectively separate sperm containing the X and Y chromosomes from all males. Alternative methods of gender selection are also being investigated. A Welsh company, Ovasort, has identified sex-specific proteins on the surface of sperm and developed antibodies against them. Its goal is to use antibodies to aggregate sperm that carry a specific sex chromosome, thereby enabling their elimination from the general population. The ART combination would also be important for sperm sex determination. Therefore, the rate of flow sorting can be increased by first removing dead and dying sperm from the population, for example by DGC or SLC. This combination can increase the "selectivity" of sperm samples. Flow sorting in IVF can produce sufficient numbers of sexual sperm, thereby creating embryos or blastocysts for further transfer. However, if flow cytometry is to become useful for species other than cattle, methods to speed up the breeding process are required to achieve the greater progeny.

2.3.3.5 Sperm Cryopreservation

Sperm cryopreservation is one of the methods to store the rare breeds and useful breeds for future breeding endeavors. The cryopreservation includes the liquid nitrogen vapors to preserve the semen for longer duration by deep freezing. As previously mentioned, the ability of cryopreserved sperm to maintain their fertilization capacity varies significantly between species. To solve this problem, new cryogenic extensions and protocols are constantly being developed. A more recent advancement has been the introduction of dimethyl sulfoxide and amides such as formamide and dimethylformamide as cryoprotectants in place of glycerin. These molecules appear to work better than glycerin in some individuals whose sperm do not freeze well, such as some stallions. One explanation for this observation is that these molecules are smaller than glycerin and therefore can cause less damage when penetrating the sperm membrane. However, no method appears to be effective for a single species. When it comes to turkey semen, developing an effective freezing method appears to require more than just new cryoprotectants and additives for the improvement in the field of poultry for better efficiency.

2.3.3.6 Removal of Viruses from Ejaculates

Viral infections are a serious issue in AI. Due to the lack of cleanliness and hygiene of animals, they get infected very soon and also to the natural mating or the physical contact with other infected animals beside them. Viral infectivity can be eliminated from the sperm of patients with viral infections such as HIV and hepatitis through a sequential sperm preparation method, that is, DGC followed by "continuous flow". Sperm from virus-infected men obtained using this method has been used in assisted reproduction studies, apparently without seroconversion of mothers and children. However, some HIV studies indicate that DGC alone cannot eliminate all infectivity of the virus. Because sperm can act as viral vectors, further work is needed to investigate the extent to which different virus particles are associated with the sperm membrane and are likely to be transmitted during treatment. In preliminary studies, the dual treatment method was also effective in eliminating equine arteritis virus from the ejaculate of infected stallion (Morrell et al., 1988). SLC and swimming have been used to reduce the infectivity of centovirus 2-enriched boar sperm virus (Blomqvist et al., 2011). Therefore, it is very important to eliminate viral infections in animals used for AI to ensure healthy, disease-free offspring.

2.4 AI IN CONSERVATION BIOLOGY

It has been suggested that AI and other forms of ART could be useful in genetic conservation and the conservation of rare breeds and endangered species. Many of these technologies have brought success at the research level, but none have produced sufficient results to improve genetic management at the population level. Sperm cryopreservation is the most widely used ART method in this regard, but most sperm frozen in so-called gene banks have never been tested for fertility. The lack of suitable females or a lack of knowledge about a species' reproductive biology may contribute to this deficit. However, long-term storage of frozen gametes of unknown fertility is not a sustainable strategy to protect rare breeds and endangered species. Development of methods for in vitro sperm fertility testing would significantly contribute to conservation efforts. Because the quality of sperm from these animals can be low, techniques such as SLC of samples prior to AI can provide significant advantages for conservation selection. AI in conservation biology mainly requires fresh semen compared to stored semen, and also it has to be provided with a suitable environment. AI cannot be done in every geographical location as it requires controlled environment for the best results.

BIBLIOGRAPHY

Bakst, M. R., & Dymond, J. S. (2013). Artificial insemination in poultry. In: M. Manafi (Ed.), *Artificial Insemination in Farm Animals*. IntechOpen, London. doi:10.5772/54918.

BBC History. (2015). *Robert Bakewell (1725–1795)*. BBC Historic Figures. https://www.bbc.co.uk/history/historic_figures/bakewell_robert.shtml

Bekana, M. (1991). *Farm Animal Obstetrics* (Monograph). Faculty of Veterinary Medicine, Addis Ababa University, Addis Ababa.

Blomqvist, G., Persson, M., Wallgren, M., Wallgren, P., & Morrell, J. M. (2011). Removal of virus from boar semen spiked with porcine circovirus type 2. *Anim. Reprod. Sci.*, 126(1–4), 108–114.

Dalton, J. C., Ahmadzadeh, A., Shafii, B., Price, W. J., & Dejarnette, J. M. (2004). Effect of simultaneous thawing of multiple 0.5 mL straws of semen and sequence of insemination on conception rate in dairy cattle. *J. Dairy Sci.*, 87(4), 972–975.

Dejarnette, J. M., & Marshall, C. E. (2005). Straw-thawing method interacts with sire and extender to influence sperm motility and conception rates of dairy cows. *J. Dairy Sci.*, 88(11), 3868–3875.

Gaurang Kumar Patel, G. K., Haque, N., Madhavatar, M., Chaudhari, A. K., Patel, D. K., Bhalakiya, N., Jamnesha, N., Patel, P., & Kumar, R. (2017). Artificial insemination: A tool to improve livestock productivity. *J. Pharmacogn. Phytochem.*, 6, 307–313.

Gibbons, A. et al. (2019). Technical recommendations for artificial insemination in sheep. *Anim. Reprod.*, 16(4), 803–809. doi:10.21451/1984-3143-AR2018-0129

Hopkins, S. S., & Evans, L. E. (1989). Artificial insemination. In: L. E. MacDonald & M. H. Pineda (Eds.), *Veterinary Endocrinology and Reproduction* (4th ed., pp. 354–388). Lea & Febiger, Philadelphia, PA.

Johnson, L. A., Flook, J. P., & Look, M. V. (1989). Flow cytometry of X and Y chromosome-bearing sperm for DNA content: Detection and separation for artificial insemination. *Gamete Res.*, 24(3), 443–452.

Kaproth, M. T., Rycroft, H. E., Gilbert, G. R., Abdel-Azim, G., Putnam, B. F., & Schnell, S. A. (2005). Effect of semen thaw method on conception rate in four large commercial dairy heifer herds. *Theriogenology*, 63, 2535–2549.

Kukovics, S., Gyoker, E., Nemeth, T., & Gergatz, E. (2019). Artificial insemination of sheep—Possibilities, realities and techniques at the farm level. In: M. Manafi (Ed.), *Artificial Insemination in Farm Animals*. IntechOpen, London. doi:10.5772/16642.

Leboeuf, B., Delgadillo, J. A., Manfredi, E., Piacere, A., Clement, V., Martin, P., Pellicer, M., Boue, P., & de Cremoux, R. (2008). Management of goat reproduction and insemination for genetic improvement in France. *Reprod. Domest. Anim.*, 43(2), 379–385.

Maes, D., Lopez Rodriguez, A., Rijsselaere, T., Vyt, P., & Van Soom, A. (2011). Artificial insemination in pigs. In: M. Manafi (Ed.), *Artificial Insemination in Farm Animals*. IntechOpen, London. doi:10.5772/16592.

Merck. (1983). *The Merck Veterinary Manual* (5th ed.). Merck, Rahway, NJ.

Morrell, J. M. (2006). Update on semen technologies for animal breeding. *Reprod. Domest. Anim.*, 41, 63–67.

Morrell, J. M. (2011). Artificial insemination: Current and future trends. In: M. Manafi (Ed.), *Artificial Insemination in Farm Animals*. IntechOpen, London. doi:10.5772/17943.

Morrell, J. M., Keeler, K. D., Noakes, D. E., & Mackenzie, N. M. (1988). Sexing of sperm by flow cytometry. *Vet. Rec.,* 122(13), 322–324.

Morrell, J. M., & Rodriguez-Martinez, H. (2009). Biomimetic techniques for improving sperm quality in animal breeding: A review. *Open Androl. J.*, 1, 1–9.

Morton, C. (n.d.). *Artificial Insemination in Horses*. PetMD. https://www.petmd.com/horse/artificial-insemination-horses

Nebel, R. L., & Jobst, S. M. (1998). Evaluation of systematic breeding programs for lactating dairy cows: A review. *J. Dairy Sci.*, 81, 1169–1174.

Phillips, P. H., & Lardy, H. A. (1940). A yolk-buffer pabulum for the preservation of bull semen. *J. Dairy Sci.*, 23, 399–404.

Pilling, D., Cardellino, R., Zjalic, M., Rischkowsky, B., Tempelman, K. A., & Hoffmann, I. (2007). The use of reproductive and molecular biotechnology in animal genetic resources management: A global overview. *Anim. Prod. Health Div. FAO*, 40, 1–13.

Roca, J., Vazquez, J. M., Gil, M. A., Cuello, C., Parrilla, I., & Martinez, E. A. (2006). Challenges in pig artificial insemination. *Reprod. Domest. Anim.*, 41(2), 43–53.

Salisbury, G. W., Fuller, H. K., & Willett, E. L. (1941). Preservation of bovine spermatozoa in yolk-citrate diluent and field results from its use. *J. Dairy Sci.*, 24, 905–910.

Samad, S. A. (1985). Diluters of the buffalo semen. In: R. H. Usmani and S. Karam Shah (Eds.), *Processing and Use of Buffalo Semen* (pp. 34–46). Animal Science Institute, National Agricultural Research Centre, Islamabad.

Webb, D. W. (1992). *Artificial Insemination in Dairy Cattle (DS58).* Animal Science Department, Florida Cooperative Extension Service, Institute of Food and Agricultural Sciences, University of Florida.

Wishart, G. J. (1985). Quantitation of the fertilising ability of fresh compared with frozen and thawed fowl spermatozoa. *Br. Poult. Sci.*, 26, 375–380.

3 Artificial Insemination in Animal Species
Current and Future Trends

Kapil Paul, Shweta Maheshwari, and Archana Saini

3.1 INTRODUCTION

Artificial insemination (AI) is a reproductive technology that involves manually transferring a male's semen into a female's uterus during heat (oestrus) period in order to create offspring. It is a method in which offspring are produced by facilitating the fusion of male and female gametes for increasing livestock production. AI is a common method used to achieve impregnation in domestic livestock, cattle, bees, and other animal species, and also in humans. AI is a technique developed for animal genetic improvement. According to Hafez and Hafez (2000), it is a valuable technique for rapid dispersal of favourable genes to facilitate genetic improvement in livestock (Murphy et al. 2017).

The ever-increasing human population is constantly raising the demand for improving the livestock production that in turn requires genetically improved animal species. The success of AI can be attributed to efficient methods of semen collection, evaluation and preservation, detection and control of oestrus cycle in female, and insemination technology. This is accomplished by selecting high-quality males producing abundant spermatozoa to inseminate hundreds of females. AI can be done with fresh as well as cryopreserved samples. For higher success rates, vigilant assessment of breeding time of female, and semen collection and its evaluation are required.

The first effective case of AI employing dogs was reported by Lazzaro Spallanzani in 1780. He used semen to inseminate a female dog showing signs of heat, which gave birth to three pups after 62 days. Today, he is often credited as a pioneer of modern AI. Later, he concluded that cooling and freezing do not destroy spermatozoa. In 1922, Ivanoff had studied AI in domestic animals, foxes, rabbits, and poultry. An artificial vagina (AV) was first crafted for the dog in 1914 by Amantea, which was then modified to collect semen from bulls. This equipment was important for the collection of semen for more effective results of AI. Intrauterine fixation method through cervix, reducing the number of sperms required. AI in India was done at Palace Dairy Farm Mysore and got 33 cows pregnant. Salisbury et al. (1941) formulated egg yolk citrate diluter. The first buffalo calf was born at Allahabad Agricultural Institute using AI in 1943. A brief overview of the history of AI is given in Table 3.1.

DOI: 10.1201/9781003452805-3

TABLE 3.1
A Brief Overview of the History of Artificial Insemination

Year	Description of Event
1677	Leeuwenhoek discovered spermatozoa
1780	Spallanzani inseminated a female dog successfully
1803	Spallanzani demonstrated that chilling did not kill sperms
1899	Ivanoff proposed AI research in Russia
1912	Ishikawa started research in Japan on AI
1914	Amantea devised artificial vagina for dogs
1937	Rectovaginal insemination technique established
1941	Citrate-buffered egg yolk preservation of sperms evolved

3.2 ARTIFICIAL INSEMINATION IN DIFFERENT ANIMAL SPECIES

3.2.1 Cattle

Ethiopia has the largest livestock population in Africa, and this sector contributes significantly to the national socioeconomic status of the country. However, this sector has been hampered by certain problems like diseases and their control policy, lack of good variety of species, and veterinary services (Mohammed 2018). The productivity of indigenous cattle is low, as cows do not give birth to calves before 35–53 months of age. The breeding must occur within 85–115 days after calving to maintain appropriate interval. After 20–50 days of postpartum, rebreeding should be done within a frame of 35–95 days.

Fixed-time artificial insemination (FTAI) has been extensively used in South America commercially for the last 20 years. According to Bó et al. (2018), use of oestradiol/progesterone protocols have raised the number of pregnancies per AI from 40% to 60% (Bo et al. 2013) because of its high effectiveness and low cost. AI in cattle has shown increased probability of birth of a male calf as compared to natural mating (Berry and Cromie 2007).

3.2.2 Pig

Use of AI is extensively practiced in pig industry because it is a promising tool to introduce superior genes into herds with lesser risk of diseases (Dominiek Maes et al. 2011). Netherlands has the highest fertility results in pig AI. Initially, the results were low with a litter size of 5–6 and pregnancy rate of just 60%. They were the first ones to initiate pooling of semen commercially but after the outbreak of swine flu in 1997, government banned the pooling. Later in 2000, there was increase in the use of AI worldwide with 90% reproductive rate and litter size of 11–13 piglets. Singh et al. (2017) stated a farrowing rate of 86% and a litter size of 14 piglets in India.

AI is a major breeding technique used commercially on farms to transfer desirable traits from high-indexed boars and prevent disease transmission. According to Bortolozzo et al. (2015), post-cervical and FTAI strategies have been developed to use in field to attain higher rates of fertility in swine species. This technique has the potential to enhance profits in pig farming and allows better maintenance of farm records (Singh and Mollier 2020).

3.2.3 Dog

The semen sample from stud dogs is collected manually for AI. It took longer to implement AI in dogs because of species-specific particularities. To improve kennel genetics, breeders can select pure-bred dogs from anywhere in the world without the stress associated with transport. This technique also helps to reduce the spread of sexually transmitted diseases, viz. Herpes virus infection (Farstad 2010). Unfavourable response of freezing to dog sperms was a major drawback to the initial work done, and this led to more research on dog AI (Linde Forsberg 2005).

According to Utomo et al. (2023), the Kintamani dog is a pure-bred Balinese dog in temporary category. Its average puberty was 7.5 months. Frozen semen sample of Kintamani dog was stored using coconut water and citric egg yolk extender (Puja et al. 2018, 2019). The first successful case of AI in Kintamani dog showed 25% conception rate by using freezing semen as demonstrated by Sulabda et al. (2022).

3.2.4 Sheep

AI in sheep has been practiced for over 50 years; inseminations were first achieved using fresh diluted semen. Intrauterine insemination using laparoscope has reduced the usage of frozen semen in sheep (Evans 1988). Manafi (2011) explained that the augmented use of AI has enhanced the production of better-quality wool, in terms of fibre diameter and staple length, in Hungarian Merino flock. Milk production also increased from 50 to 70 L to 300 L per ewe annually (Barillet et al. 1993).

Certain factors should be considered for satisfactory results in AI such as age of the ram, semen quality, and mass motility. In Errandonea et al.'s (2018) study, improvement in the ovulation rate and fecundity has been observed after giving short-term protein supplementation before AI using fresh semen. High- and low-fertility rams made a difference in sperm quality parameters affecting the success rates of AI (Vicente-Fiel et al. 2014). The success rate depends not only upon male but also on the fertility of female, type of oestrus, and season. However, pregnancy rates exceeded 80%, and 90% in a few exceptional cases.

3.3 ADVANTAGES

AI offers many potential advantages over natural breeding but the most important was to control venereal diseases that spread due to direct contact between male and female. Later, genetic improvement of food-producing animals became another important reason to promote AI. It is used as means of semen evaluation and monitoring, storage of semen for future, routine examination of female reproductive system, injury risk reduction, improvement of less fertile species, and allowing fixed-time breeding.

A male deposits much more semen in female during natural mating than is required to produce a pregnancy. Moreover, natural breeding is physically more stressful and can cause injury to either male or female. The collected semen can be used in diluted form and single ejaculate can produce hundreds of doses, hence allowing impregnation of multiple females at the same time but at different geographical sites, thereby reducing the number of males required for breeding.

In developing countries, AI plays important role in boosting animal productivity, especially for milk and wool. Valergakis et al. (2007) indicated that daughters of AI sires were producing almost 900kg of extra milk per lactation than daughter of normal bulls. Another advantage is the reduction in the number of males that farmers need to maintain on the farm, thereby eliminating the cost of accommodation and danger/injury to farm staff. The significant cost of housing and labour is involved in keeping such animals that can be avoided by the use of AI.

Natural breeding leads to the transmission of venereal diseases, which can be controlled by using AI. Some diseases/pathogens can spread through semen but collection process allows screening of such disease agents. Semen of desired male can be used even after the death of the animal. Chatikobo (2009) described that collected and stored semen samples are regularly examined for quality, thereby eluding complications related to male infertility. Mating of animals with big differences in size and geographical barriers can be achieved without injury to any animal. Genetic improvement of the species can be achieved by the selection of pure bred and high-quality bred used in AI. It is a technique used to upgrade unimproved animals in remote areas. AI helps to facilitate international livestock trade by importing high-level semen rather than moving animals themselves. Therefore, problems related to acclimatization can be eliminated as well.

3.4 PROCEDURE OF ARTIFICIAL INSEMINATION

For the process of AI, healthy male cattle is identified. Semen is collected by a skilful person by using sterile non-toxic equipment, because spermatozoa are highly sensitive to toxic materials. Toxic effects can be seen in fraction of seconds of contact with any kind of toxic material. Semen is then diluted with extenders and antibiotics are added. Semen is then either preserved or used for insemination at the same time. A female cattle in heat period is identified, and insemination is carried out in various ways depending upon the species. An outline of AI procedure is shown in Figure 3.1 (Brinsko and Varner 1992).

3.4.1 Collection and Constituents of Semen

A disease-free healthy male is identified by professional and qualified veterinarians for any infectious agents prior to selecting it as semen donor. Health is thoroughly checked, and they are vaccinated for bovine viral diarrhoea, infectious bovine rhinotracheitis, and trichomoniasis (Hansen 2006), because a bull not only donates spermatozoa to produce offspring but important features like genetic potential are also passed on to offspring, which is far more important than average milk or meat production (Herman et al. 1994). After the assessment, potential candidate for semen collection is identified and selected. For the collection of semen, a non-toxic sterile AV is used, which is a 45-cm long rubber barrel with rough inner lining coated with sterile and non-spermiotoxic lubricant. After the ejaculation, the volume of

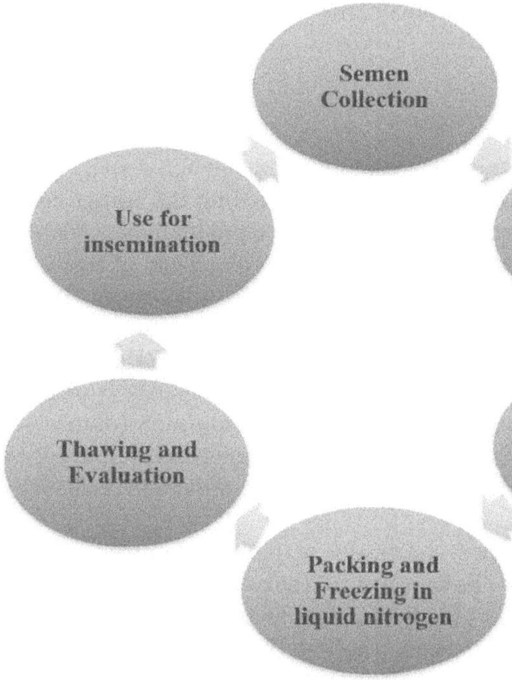

FIGURE 3.1 Outline of the procedure of artificial insemination.

the semen is measured by putting into graduated vial. Semen is then studied under microscope to find the motility and morphology of viable sperms.

3.4.2 Processing of Semen

According to a study by Cseh et al. (2012), semen should be stored carefully after collection because higher temperatures can destroy sperm. As a result, immediately after semen collection, extenders like milk, glycerine, egg yolks, etc. are added to the semen to dilute it and give the sperms nutrition while it is being stored.

3.5 PRESERVATION OF SEMEN

After the collection of semen, percentage of motile spermatozoa is assessed. If the semen is to be used immediately, appropriate extender is added to the semen to extend the viability of spermatozoa (Brinsko and Varner 1992). But if spermatozoa in semen are in higher percentage or the semen is to be used later, then it is necessary to store it in controlled temperature conditions to ensure the good health of the spermatozoa at the time of insemination. In properly sterilized AV, it is necessarWWy to maintain temperature of ejaculated semen near the body temperature (37°C–39°C) and light should be avoided during the transportation to laboratory for further processing. A suitable extender should be mixed to semen within few minutes of collection of semen. Quality of ejaculated semen should be screened properly to get good results from it (Brinsko and Varner 1992).

3.5.1 Liquid Preservation of Semen

To store the semen in liquid form to maintain the metabolic activity of spermatozoa. It should be protected from cold shock by avoiding the abrupt cooling from body temperature to 5°C. So semen should be cooled gradually to avoid cold shock. However, even with gradual cooling, it cannot be completely avoided (Parkinson and Morrell 2019). Cell membranes are damaged by cold shock, leading to its membrane become more permeable, and intracellular potassium, enzymes, lipids, cholesterol, lipoprotein, and adenosine triphosphate seep out (Hossain et al. 2022). Semen is preserved from cold shock by acquiring different means from time to time. The addition of egg yolk or milk in the diluents is very useful. So diluent containing 20% of egg yolk and a buffer is considered the most beneficial. While egg yolk is by far the most often utilized addition, skim milk, whole milk, and coconut milk have also been used successfully.

Phosphocaseinate is developed chemically to replace skimmed milk extenders (Batellier et al. 2001). Phosphate buffer served as the basis for early diluents, but citrate quickly outperformed them, most likely due to its ability to increase the solubility of egg yolk proteins. Zwitterion buffers such as TES (N-tris (hydroxymethyl) methyl-2-aminoethane sulphonic acid) or TEST (Tris titrated with TES) are also widely used with great success. It has been proposed that Bovine Semen Plasma (BSP) is bound by low density lipoproteins (LDL), namely phosphatidylcholine, in the extension, preventing it from interacting with the spermatozoa and thereby shielding the membranes from harm. It has been suggested that BSP is bound by LDL, namely phosphatidylcholine, in the extension, preventing it from interacting with the spermatozoa and thereby protecting the membranes from harm (Holt 2000). It is suggested that LDL, particularly phosphatidylcholine, binds BSP in the extender, preventing its interaction with the spermatozoa and hence protecting the membranes from damage (Vishwanath and Shannon 2000). According to Evans and Maxwell (1987), the fertility of ram semen only lasts for 12–24 hours, but that of bovine semen kept at 5°C in such diluents lasts for 2–4 days (Hafs 1961). Semen of horse and dogs is frequently stored at 5°C for short term, as it does not show any response to freezing (Parkinson and Morrell 2019).

3.5.2 Cryopreservation

Storage of semen for a longer period of time is carried out by cryopreservation, which helps in the use of desired genetic material across the globe. For preservation of semen by this method, semen is examined to assess for any infection. Fertility life of spermatozoa is maintained for longer span of time. Various chemicals such as glycerol are used as cryoprotectants to prevent the spermatozoa from cold shock (Polge et al. 1949) and serious consequences of ice crystal formation. This ice crystal is prevented by dehydration at slower rate, which leads to good preservation of spermatozoa in the semen. Various new processes are derived to prevent spermatozoa from ice crystal formation, such as vitrification, a process that solidifies a liquid without forming ice crystals. According to recent research, vitrification of spermatozoa can occur through rapid freezing without the need for a cryoprotectant. Thawing is also a critical procedure involved in AI process. It should be done at faster rate, because thawing at slower rate leads to recrystallization of ice crystals inside the cell to damage the cell membrane. For Bovine semen, it is suggested to be done at 35°C–38°C for 40 s. Sometimes it is done at 60°C for shorter duration to improve the rate of recovery (Rauthan and Negi 2022).

3.6 OESTRUS AND OVULATION DETECTION

Oestrus is defined as a time when the female exhibits typical sexual behaviours, such as immobility, arching the back or raising the hindquarters, and pricking of the ears—features that together are referred to as lordosis in small laboratory animals. Females also frequently engage in riding and mounting behaviours (Ahmed 2018). In artificially breeding herds, detection of oestrus is essential for successful reproduction. Higher numbers of cattle in a herd make it harder to visually monitor each animal, decreasing estrus detection rates, while high milk production requires more energy, which can lead to reduced expression of estrus signs, further complicating detection (Ahmed 2018; Rauthan and Negi 2022). Da Costa et al. (2011) suggested that as milk production rises, the efficacy of heat detection and the number of conceptions per breeding are also increased. This suggests that producers can have both superior reproductive performance and a higher milk output. A process that uses AI to take advantage of the superior genetics is encouraged by effective heat detection (Ahmed 2018).

3.6.1 Natural and Artificial Ovulation Induction

Cattle's oestrus cycle, which normally lasts 21 days (varying from 17 to 24 days) for nonpregnant cows, is a naturally occurring, cyclical process that primes the reproductive system for oestrus (heat) and ovulation. The four distinct phases of this cycle are proestrus (progesterone levels fall as the corpus luteum regresses, and a new follicle begins to form), oestrus (period of sexual receptivity, characterized by cow standing to be mounted by a bull or another female; oestrus typically lasts 6–30 hours, with an average of 20 hours), metestrus (phase after oestrus during which ovulation occurs), and diestrus (longest phase when corpus luteum forms and produces progesterone). Cattle's oestrus cycle is a normal and vital aspect of reproduction during which the reproductive tract to go through cyclic change. Hormones like oestrogen and progesterone control the growth of the reproductive tract and prepare it for possible fertilization and guaranteeing the ongoing chances for mating and pregnancy (Forde et al. 2011; Rauthan and Negi 2022).

In cattle, artificial induction of ovulation is the process of injecting hormones to cause the ovaries to release eggs, usually in order to increase fertility and allow timed artificial insemination. Either direct use of hormones that imitate the actions of luteinizing hormone (LH) or stimulation of the pituitary gland to release LH. Gonadotropin-releasing hormone stimulates the pituitary gland to release LH, which in turn triggers ovulation. These hormones are injected for the stimulation of ovaries to release eggs (Laplacette et al. 2025; Wiltbank and Pursley 2014; de Medeiros Bastos et al. 2004).

3.6.2 Deposition of Semen in the Female

Semen containing spermatozoa are deposited in different sites of reproductive tract as the location of semen deposition during natural mating varies among animals. Semen is deposited intrauterine in case of horses, dogs, and camels, whereas deposition is done in vagina in case of ruminants and primates. Semen can be delivered in the uterus in the majority of animals by inserting an insemination catheter via the cervix but the cervix's tightly folded shape makes it difficult for an insemination catheter to pass through, in case of sheep and goats. Smaller volume of semen can be delivered into uterus, as it already bypassed the barrier (cervix). Volume of semen varies in reference to site of deposition; lesser dose of semen is required in case of delivery into uterus as compared to cervix. $10–20 \times 10^6$ spermatozoa per dose is considered as good number to induce pregnancy in cattle.

3.7 FUTURE OF ARTIFICIAL INSEMINATION

Numerous new technologies have been discovered for getting higher success rates in the production of individuals of particular species. These techniques facilitate the farm owners to produce higher number of individuals leading to get more profits. These developments increase the probability of pregnancies, enable offspring genetic selection, and improve herd management. Some of these advancements are explained ahead.

3.7.1 Centrifugation

In order to improve the quality of the semen and lower the volume required for insemination, centrifuges are used in AI for cattle. This enhances reproductive results, particularly when working with lower-quality semen samples. Centrifugation facilitates in using lesser quantity of semen samples and increase the success rate up to 70%–80% (Gloria et al. 2016).

3.7.2 Sex Selection

Separation of X-chromosome-bearing sperm from the semen sample to produce female calf is termed as Sexed semen. Hence, it leads to predetermination of offspring sex, useful for dairy cattle (more female calves) and beef cattle (more male calves for meat production). Female participation in breeding projects has been made easier by genomic selection in conjunction with both well-established and newly developed assisted reproductive technologies. In order to significantly reduce generation intervals, these technologies are crucial (Ayantoye et al. 2025; Mikkola et al. 2024).

3.7.3 Removal of Pathogens from Semen

During initial days, AI was vulnerable to transmission of diseases from infected semen to female cattle (Maes et al. 2008). To eliminate these pathogens, suitable antibiotics are mixed in seminal extenders during the collection of semen. Some commonly used antibiotics are Polymyxin B sulphate, crystalline penicillin, gentamycin sulphate, amikacin sulphate, and ticarcillin which are added to seminal extenders in different concentrations.

3.8 DRAWBACKS OF ARTIFICIAL INSEMINATION

Some disadvantages are a part of AI. As ejaculated spermatozoa are prone to environmental injuries, it requires a specialist or skilful person to perform it. The cost of purchasing the supplies and equipment required for AI may raise the breeding program's overhead expenses. Cost of instruments required for transportation and storage of semen is also very high. There is always a risk of human injury during the collection of semen with AV. So it is necessary to have a trained person to collect semen (Esmaeili Dahesht et al. 2010; Parkinson and Morrell 2018).

3.9 CONCLUSION

The livestock industry has much to gain from AI due to its broad applicability to reproductive management systems. AI is considered one of the most effective animal reproductive technologies since it provides maximum development of advanced sires for use over large herds AI decreases the chances of transmitting infections that can be spread through natural mating, or assists in controlling breeding time when used in conjunction with various birth scheduling techniques to enhance the herd's productivity. Nonetheless, successful AI implementation requires proper management of personnel skilled in animal reproduction, attention and resources, and effective heat detection along with timely decision-making. If all requirements are met, AI serves as a great enhancement tool to the best genetic improvement programs for mass livestock production.

REFERENCES

Ahmed, Mohammed. 2018. Artificial Insemination and Its Economical Significancy in Dairy Cattle: Review. *International Journal of Research Studies in Microbiology and Biotechnology,* 4(1): 30–43.

Ayantoye, Jesse Oluwaseun, Hubdar Ali Kolachi, Xiaomeng Zhang, Muhammad Shahzad, Omaima Mohamed Tawfik Kandil, Pengcheng Wan, and Xueming Zhao. 2025. Advances in Timed Artificial Insemination: Integrating Omics Technologies for Enhanced Reproductive Efficiency in Dairy Cattle. *Animals,* 15(6): 816.

Barillet, Francis, S. Sanna, Didier Boichard, J. M. Astruc, A. Carta, and S. Casu. 1993. Genetic evaluation of the Lacaune, Manech and Sarda Dairy Sheep with an Animal Model. *Journal of Animal Production*, 1(Suppl. 1): 580–607.

Batellier, F, M Vidament, J Fauquant, G Duchamp, G Arnaud, J.M Yvon, and M Magistrini. 2001. Advances in Cooled Semen Technology. *Animal Reproduction Science,* 68(3–4): 181–190.

Berry, D. P., and A. R. Cromie. 2007. Artificial Insemination Increases the Probability of a Male Calf in Dairy and Beef Cattle. *Theriogenology,* 67(2): 346–352.

Bó, Gabriel A., Emilio Huguenine, José Javier de la Mata, Richard Núñez-Olivera, Pietro S. Baruselli, and Alejo Menchaca. 2018. Programs for Fixed-Time Artificial Insemination in South American Beef Cattle. *Animal Reproduction,* 15(Suppl 1): 952.

Bortolozzo, F. P., M. B. Menegat, A. P. G. Mellagi, M. L. Bernardi, and I. Wentz. 2015. New Artificial Insemination Technologies for Swine. *Reproduction in Domestic Animals,* 50: 80–84.

Brinsko, Steven P, and Dickson D Varner. 1992. Artificial Insemination and Preservation of Semen Artificial Insemination History. *Veterinary Clinics of North America: Equine Practice,* 8 (1): 205–248.

Chatikobo, Paul. 2009. Artificial Insemination in Rwanda. *Dairy Mail Africa: Publication for the Dairy Industry in Africa,* 4(3): 18–23.

Costa, Antonio Nelson Lima, Airton Alencar Araujo, and Jose Valmir. 2011. Particularities of Bovine Artificial Insemination. In: M. Manafi (Ed.), *Artificial Insemination in Farm Animals.* InTech, Rijeka, Croatia.

Cseh, S., V. Faigl, and G. S. Amiridis. 2012. Semen Processing and Artificial Insemination in Health Management of Small Ruminants. *Animal Reproduction Science,* 130(3–4): 187–192.

de Medeiros Bastos, Guilherme, Ricardo Heitor Brenner, Fabŕicio Wollman Willke, Jairo Pereira Neves, João Francisco Coelho de Oliveira, José Francisco Manta Bragança, Sérgio Abreu Machado, Patŕicia Marafon Porciúncula, and Paulo Bayard Dias Gonçalves. 2004. Hormonal Induction of Ovulation and Artificial Insemination in Suckled Beef Cows under Nutritional Stress. *Theriogenology,* 62(5): 847–853.

Errandonea, N., S. Fierro, C. Viñoles, J. Gil, G. Banchero, and J. Olivera-Muzante. 2018. Short Term Protein Supplementation during a Long Interval Prostaglandin-based Protocol for Timed AI in Sheep. *Theriogenology,* 114: 34–39.

Esmaeili Dahesht, L., H. Negarestan, A. Eimanifar, F. Mohebbi, and R. Ahmadi. 2010. The Fluctuations of Physicochemical Factors and Phytoplankton Populations of Urmia Lake, Iran. *Iranian Journal of Fisheries Sciences,* 9(3): 368–381.

Evans, G. 1988. Current topics in Artificial Insemination of Sheep. *Australian Journal of Biological Sciences,* 41(1): 103–116.

Evans, G, and W. M. C. Maxwell. 1987. Salamons' Artificial Insemination of Sheep and Goats. In: G. Evans and W. M. C. Maxwell (Eds.), *Salamons' Artificial Insemination of Sheep and Goats* (2nd ed., p. 194). Butterworths, Sydney, Australia.

Farstad, Wenche K. 2010. Artificial Insemination in Dogs. In: England, G. and von Heimendahl, A. (Eds.), *BSAVA Manual of Canine and Feline Reproduction and Neonatology* (pp. 80–88). BSAVA Library, Gloucester, UK.

Forde, N., M.E. Beltman, P. Lonergan, M. Diskin, J.F. Roche, and M.A. Crowe. 2011. Oestrus Cycles in Bos Taurus Cattle. *Animal Reproduction Science,* 124(3–4): 163–169.

Gloria, Alessia, Augusto Carluccio, Laura Wegher, Domenico Robbe, Giovanni Befacchia, and Alberto Contri. 2016. Single and Double Layer Centrifugation Improve the Quality of Cryopreserved Bovine Sperm from Poor Quality Ejaculates. *Journal of Animal Science and Biotechnology,* 7(1): 30.

Hafs, H. D. 1961. Fertility of Bull Sperm with Added Catalase. *Journal of Dairy Science,* 44(8): 1529–1536.

Hafez, B., and Hafez, E.S.E. 2000. Anatomy of Female Reproduction. In: B. Hafez and E.S.E. Hafez (Eds.), *Reproduction in Farm Animals.* doi: 10.1002/9781119265306.ch2.

Hansen, G.R. 2006. *Managing Bull Fertility in Beef Cattle Herds.* Animal Science Department. Florida Cooperative, Extension, Service, Institute of Food and Agricultural Sciences, University of Florida.

Herman H.A., Mitchell J.R., Doak G.A. 1994. *The Artificial Insemination and Embryo Transfer of Dairy and Beef Cattle: A Handbook and Laboratory Manual.* IPP, Interstate Publishers, INC., Danville, IL, pp. 3–7; 45–244.

Holt, W.V. 2000. Basic Aspects of Frozen Storage of Semen. *Animal Reproduction Science,* 62(1–3): 3–22.

Hossain, Sheikh Mohammad Jahangir, Md. Faizul Hossain Miraz, Shahrina Akter, Md. Younus Ali, and Gautam Kumar Deb. 2022. Comparative Study on Semen Quality and Fertility of Red Chittagong Cattle, BCB1 and Munshiganj Bulls of Bangladesh. *Advances in Bioscience and Biotechnology,* 13(9): 428–441.

Ivanoff, E. I. 1922. On the Use of Artificial Insemination for Zootechnical Purposes in Russia. *The Journal of Agricultural Science,* 12(3): 244–256.

Laplacette, A. L., C. Rial, E. Sitko, M. M. Perez, S. Tompkins, M. L. Stangaferro, M. J. Thomas, and J. O. Giordano. 2025. Delaying Induction of Ovulation and Timed Artificial Insemination in a Double-Ovsynch Protocol Increased Expression of Estrus and Altered First-Service Reproductive Outcomes of Lactating Dairy Cows. *Journal of Dairy Science,* 108(1): 1103–1124.

Linde-Forsberg, C. 2005. *Recent Advances in Small Animal Reproduction.* International Veterinary, Ithaca.

Maes, Dominiek, Alfonso Lopez Rodriguez, Tom Rijsselaere, Philip Vyt, and Ann Van Soom. 2011. "Artificial Insemination in Pigs." In *Artificial Insemination in Farm Animals.* InTech. doi:10.5772/16592.

Maes, D., H. Nauwynck, T. Rijsselaere, B. Mateusen, P. Vyt, A. de Kruif, and A. Van Soom. 2008. Diseases in Swine Transmitted by Artificial Insemination: An Overview. *Theriogenology,* 70(8): 1337–1345.

Manafi, Milad (Ed.). 2011. *Artificial Insemination in Farm Animals.* InTech, Rijeka, Croatia.

Mikkola, Marja, Karolien Leen Jan Desmet, Elisabeth Kommisrud, and Michael A. Riegler. 2024. Recent Advancements to Increase Success in Assisted Reproductive Technologies in Cattle. *Animal Reproduction,* 21(3): e20240031.

Mohammed, Ahmed. 2018. Artificial Insemination and its Economical Significance in Dairy Cattle. *International Journal of Research Studies in Microbiology and Biotechnology,* 4(1): 30–43.

Murphy, Edel M., Craig Murphy, Ciara O'Meara, Gemma Dunne, Bernard Eivers, Patrick Lonergan, and Sean Fair. 2017. A Comparison of Semen Diluents on the In Vitro and In Vivo Fertility of Liquid Bull Semen. *Journal of Dairy Science,* 100(2): 1541–1554.

Parkinson, Timothy J., and Jane M. Morrell. 2019. Advantages and Disadvantages of Artificial Insemination. In: M. Manafi (Ed.), *Veterinary Reproduction and Obstetrics* (10th ed., pp. 746–777). WB Saunders, Philadelphia, PA.

Polge, C, A. U. Smith, and A. S. Parkes. 1949. Revival of Spermatozoa after Vitrification and Dehydration at Low Temperatures. *Nature,* 164(4172): 666.

Puja, I. K., N. M. Sawitri, N. Maharani, I. W. N. F. Gunawan, and L. G. S. S. Heryani. 2018. A Comparative Study on the Effects of Coconut Water based Extenders on the Quality of Kintamani Dog Semen Preserved at 40°C. *Advances in Animal and Veterinary Sciences,* 6(5): 192–196.

Puja, I. Ketut, Ni Made Sawitri, Nisa Maharani, Luh Gde Sri Surya Heryani, Anak Agung Gde Oka Dharmayudha, and I. Wayan Nico Fajar Gunawan. 2019. Preservation of Semen from Kintamani Bali Dogs by Freezing Method. *Journal of Advanced Veterinary and Animal Research,* 6(2): 158.

Rauthan, Aishee, and Ankur Negi. 2022. Artificial Insemination in Cattle. *International Journal of Veterinary Sciences and Animal Husbandry,* 7(1): 5–8.

Salisbury, G. W., H. K. Fuller, and E. L. Willett. 1941. Preservation of Bovine Spermatozoa in Yolk-Citrate Diluent and Field Results from its Use. *Journal of Dairy Science,* 24(11): 905–910.

Singh, M., R. T. Mollier, and D. J. Rajkhowa. 2017. Mega Seed Project on Pig and Poultry Seed Project in the Service of Tribal Farmer's of Nagaland for Livelihood and Income Enhancement. *Souvenir for North Eastern Zone Regional Agri-Fair,* 18: 36–39.

Singh, Mahak, and R. Talimoa Mollier. 2020. Artificial Insemination in Pig, Its Status and Future Perspective in India: A Review. *The Indian Journal of Animal Sciences,* 90(9): 1207–1212.

Sulabda, I. N., A. A. G. O. Dharmayudha, I. W. N. F. Gunawan, D. F. Hashilonda, L. G. Setyawati, and I. K. Puja. 2023. First Successful Pregnancy Following Intravaginal Insemination Using Frozen Semen in Kintamani Dog. *Advances in Animal and Veterinary Sciences*, 11(5): 720–724.

Sulabda, I. Nyoman, Anak Agung Gde Oka Dharmayudha, and I. Ketut Puja. 2022. Comparative Analysis of One-Step and Two-Step Dilution on Quality of Frozen Semen in Kintamani Dogs. *World's Veterinary Journal,* 3: 284–289.

Utomo, Bravanasta Glory Rahmadyasti, I. Ketut Puja, and I. Wayan Nico Fajar Gunawan. 2023. Dynamic of Vaginal pH and Ovary Ultrasound Imaging of Kintamani Bali Bitch during Proestrus to Estrus Phase. *International Journal of Veterinary* Science, 12(2): 242–247.

Valergakis, G. E., G. Arsenos, and G. Banos. 2007. Comparison of Artificial Insemination and Natural Service Cost Effectiveness in Dairy Cattle. *Animal,* 1(2): 293–300.

Vicente-Fiel, S., I. Palacín, P. Santolaria, E. Fantova, F. J. Quintín-Casorrán, E. Sevilla-Mur, and J. L. Yániz. 2014. In vitro Assessment of Sperm Quality from Rams of High and Low Field Fertility. *Animal Reproduction Science,* 146(1–2): 15–20.

Vishwanath, R., and P. Shannon. 2000. Storage of Bovine Semen in Liquid and Frozen State. *Animal Reproduction Science,* 62(1–3): 23–53.

Wiltbank, Milo C, and J Richard Pursley. 2014. The Cow as an Induced Ovulator: Timed AI after Synchronization of Ovulation. *Theriogenology,* 81(1): 170–185.

4 Livestock at the Crossroads

Sustainability in a Global Agri-Food Landscape through Application of Biotechnology

Rekha Dhanai and Surbhi Pradhan

4.1 INTRODUCTION

Livestock refers to domesticated, terrestrial animals that are raised in agricultural settings for the purpose of providing traction or producing goods such as milk, meat, eggs, and hides. They play a variety of roles for different populations, contributing to distinct agri-food systems around the world. Livestock development, food and nutritional security, animal welfare and health, and ethical and environmental issues are the four key and interconnected facets of the global livestock industry's sustainability.

4.2 LIVESTOCK DEVELOPMENT THROUGH BIOTECHNOLOGY

Animal biotechnology is a very useful method of altering living things when combined with science and engineering. Transgenic animals are the best illustration of animal biotechnology. The output of livestock in India was anticipated to peak in 2020–2021 and then rise quickly by 2027 (Report: Indian Animal Feed Market, 2022–2027). The development of products, the improvement of animals, and the production of microorganisms for specific agricultural applications are the main objectives of animal biotechnology. In order to fulfill the demand, livestock output was increased throughout the past year using conventional livestock development techniques (Chander and Prakashkumar, 2013). Biotechnology refers to the application of artificial methods to modify genetic material, enabling the production of novel compounds through selective breeding.

Using biotechnology, genetic engineering allows us to directly alter an organism's genes. It is a collection of technologies that alter cells' genetic composition, including transferring genes both within and between species to create new or improved organisms. Biotechnological methods can also be used to evaluate the ideal diet by evaluating the gene expression for a particular attribute, such as the production of

DOI: 10.1201/9781003452805-4

meat and milk (Bimrew, 2014). The three types of biotechnology that impact animal health, nutrition, breeding, and reproduction are biological, chemical, and physical techniques (Salem et al., 2007). Biotechnology plays a specific role in the production of livestock. Through increased growth, better carcass quality and reproduction, better nutrition and feed utilization, better quality and safety of animal feed, improved animal health and welfare, and reduced waste due to more effective resource use, it will offer new and surprising opportunities to increase animal productivity (Bimrew, 2013).

Cloning and cryopreservation are methods that clone animals with superior features and are useful in the conservation of endangered species. By producing an identical genetic duplicate of an animal, cloning enables the reproduction of genetically better species. For breeding purposes, this has been employed to duplicate desirable animals with exceptional qualities, such as meat-producing livestock or high-yield dairy cows (Hur, 2017). Freezing sperm, eggs, or embryos for later use is known as cryopreservation. This ensures that valuable genetic features can be passed on to future generations even after the death of high-value animals by preserving their genetic material (FAO, 2004). This method is especially useful in agriculture and the conservation of endangered animal species (Table 4.1).

4.3 BIOTECHNOLOGY IN ANIMAL NUTRITION

There is not enough area left over to raise fodder because more land is being exploited for agricultural purposes. The application of various chemicals, like feed additives, to enhance the nutritional content and the value of animal feeds as well as the digestibility of inferior feeds, especially roughage, is the primary way that biotechnology contributes to the production of livestock (McSweeney et al., 2019). Biotechnology can improve animal's nutritional value by protecting fat, protein, and amino acids (Shete, 2023). In comparison to the energy diet and other dietary treatments, the combination of antibiotic and enzymes led to a much-reduced concentration of *Escherichia coli* in small intestines (Sarica et al., 2005). Biotechnology has significantly enhanced animal productivity, feed quality (fat, protein, and amino acid preservation; decrease in anti-nutritional substances, etc.), and animal nutrient availability (via the use of enzymes, prebiotics, probiotics, etc.). A growth in feed production is required to meet this increase in output. To a certain degree, biotechnology can assist with these problems.

4.4 HEALTH AND ANIMAL WELFARE

Numerous applications of biotechnology are available to improve the well-being and health of animals. Among the main applications of biotechnology in animal health and welfare are as follows:

i. **Gene Therapy**: This therapy treats or prevents genetic diseases in animals. Through the insertion, modification, or removal of genes within a patient's cells, biotechnology makes it possible to treat genetic illnesses (Sheridan, 2011). Conditions including sickle cell anemia, cystic fibrosis, and some

TABLE 4.1
Advancements in Agriculture Sector through Biotechnology (Graff and Newcomb, 2003)

Initial Phase (1996–2002)	From 2003 Onward	Current Status
Focused on maize, soybean, and cotton	Additionally focused on tomato, tobacco, potato, etc.	Current focus of bioengineering is on drought-tolerant corn, pest-resistant cowpea, golden rice, and herbicide-tolerant crops, and animal biotechnology focuses on welfare, productivity, reproduction, and disease resistance of dairy and beef cattle, chickens, and aquaculture.
Focused on crop protection: mainly protection of a single trait	Greater focus on nutritional qualities and consumer-oriented food quality, pharmaceuticals and biomaterials, and many stacked features for crop protection.	Bt technology enables the development of insect-resistant crops, while gene editing facilitates resistance to diseases and herbicidess.
Organized farm acceptance	Suppliers collaborate closely with stakeholders in the farm sector; new crop markets are cautiously accepted.	Current focus of bioengineering for organized farms is mostly about making crops more resistant to diseases, increasing yields, and making farming more environmentally friendly. This includes making genetically modified organisms that have superior traits.
Limited utilization of segregation for GM crops	New techniques for GM crop testing and segregation may be created in the areas of production, transportation, and processing.	Genetically modified (GM) crops and animals are segregated to support research, enforce non-GM standards, and ensure customer choices.

forms of cancer may be cured or lessened as a result. It can also be used to treat hereditary illnesses in pets.

ii. **Artificial Insemination (AI)**: Controlled animal breeding is made possible by AI, which boosts genetic diversity as well as output. AI and biotechnology can be used to rapidly disseminate superior genetics throughout a population (Gupta et al., 2017).

iii. **Disease Control**: Animals that have undergone genetic alteration may become more disease-resistant. To improve animal welfare and lessen the need for medicines, certain genetically modified birds are immune to avian influenza, while cattle can be modified to be resistant to foot-and-mouth disease or mastitis (Saegerman et al., 2012). Additionally, vaccines are developed using biotechnology to improve the health of pets and animals.

iv. **Diagnostic Tools**: Advanced diagnostic methods like PCR (polymerase chain reaction), which is used to identify infections and genetic diseases, are made possible by biotechnology. Early disease detection and therapy are enhanced by these techniques (Klietmann and Ruoff, 2001).

v. **Alternatives to Animal Testing**: Animal testing in research and medication development has been replaced by biotechnology. Animal experimentation is no longer necessary thanks to organ-on-a-chip technologies and cell-based assays, which enable more moral and efficient testing (Deinsberger et al., 2020).
vi. **Improved Livestock Management**: Better living circumstances and healthier animals can be ensured by monitoring livestock for diseases, nutritional requirements, and genetic features using genetic sequencing and biotechnology. This lessens the adverse effects of livestock production on the environment and supports sustainable farming methods (McSweeney et al., 2019).
vii. **Animal Cloning and Breeding**: Cloning and sophisticated (advanced) breeding methods are made possible by biotechnology, guaranteeing superior animals for use in agriculture (Hur, 2017). These technologies can assist ensure the survival of endangered species and enhance the general health of cattle, despite their contentious nature in some places (Saegerman et al., 2012).

As a result, biotechnology plays a significant role in improving animal welfare and health by providing creative ways to enhance animal lives.

4.5 ETHICAL AND ENVIRONMENTAL CONSIDERATIONS

Although biotechnology has many advantages, there are also moral and environmental issues that must be resolved:

- **Genetic Modification**: Changing a species' natural genetic composition through genetic modification of plants and animals presents moral dilemmas (WHO, 2010). This includes worries about potential unintended repercussions and long-term ecological implications.
- **Sustainability**: Excessive use of biotechnology, including genetically modified organisms, may result in biodiversity loss and other unforeseen environmental effects.

In the end, the agriculture sector and society at large gain from farmers and researchers using biotechnology to produce animals that are healthier, more productive, and more suited to environmental difficulties. Biotechnology has significantly improved the welfare of animals. The field keeps developing, providing fresh approaches to farming, sustainable practices, and illness. The ethical and environmental implications of these developments must be carefully considered, though.

REFERENCES

Bimrew, A., 2013. Potential of biotechnology in animal feed improvement in developing countries. *Biotech Article*, 2: 15–28.

Bimrew, A., 2014. Biotechnological advances for animal nutrition and feed improvement. *World Journal of Agricultural Research*, 2(3): 115–118.

Chander, M., and Prakashkumar, R., 2013. Investment in livestock extension activities by State Departments of Animal Husbandry (SDAH) in India: An appraisal. *Indian Journal of Animal Sciences*, 83(2): 185–189.

Deinsberger, J., Reisinger, D., and Weber, B., 2020. Global trends in clinical trials involving pluripotent stem cells: A systematic multi-database analysis. *Npj Regenerative Medicine*, 5(1): 15. doi:10.1038/s41536-020-00100-4.

FAO, 2004. Conservation Strategies for Animal Genetic Resources by D.R. Notter, Background Study Paper no. 22. Commission on Genetic Resources for Food and Agriculture, Rome.

Graff, G. D., and Newcomb, J., 2003. *Agricultural Biotechnology at the Crossroads: Part I: The Changing Structure of the Industry.* Bio-Era-Bio Economic Research Associates.

Gupta, R.O., Attupuram, N.M., Saha, S., and Bhosale, V.P., 2017. Dissemination of genetics through AI in India. *Indian Dairyman*, 11: 68–74.

Hur, S. J., 2017. A study on current risk assessments and guidelines on the use of food animal products derived from cloned animals. *Food and Chemical Toxicology*, 108: 85–92. doi:10.1016/j.fct.2017.07.047.

Indian Animal Feed Market, 2022. *Industry Trends, Share, Size, Growth, Opportunity and Forecast 2022–2027.* IMARC Group.

Klietmann, W.F., and Ruoff, K.L., 2001. Bioterrorism: Implications for the clinical microbiologist. *Clinical Microbiology Reviews*, 14: 364–381.

McSweeney, C.S., Dalrymple, B.P., Gobius, K.S., Kennedy, P.M., Krause, D.O., Mackie, R.I., and Xue, G. P., 2019. The application of rumen biotechnology to improve the nutritive value of fibrous feedstuffs: Pre-and post-ingestion. *Livestock Production Science*, 59(2–3): 265–283. https://www.researchgate.net/publication/379652888_Biotechnology_in_animal_nutrition_and_feed_utilization.

Saegerman, C., Del Pozzo, F., and Humblet, M.F., 2012. Reducing hazards for humans from animals: Emerging and re-emerging zoonoses. *Italian Journal of Public Health*, 9: 13–24.

Salem, H.B. et al. 2007. Feed supplementation blocks for increased utilization of tanniniferous forages by ruminants. In: Makkar HPS, Sanchez M, Speedy AW (Eds.), *Feed Supplementation Blocks. Urea-Molasses Multinutrient Blocks: Simple and Effective Feed Supplement Technology for Ruminant Agriculture* (FAO Animal Production and Health Paper 164, pp. 1–12). FAO.

Sarica, S.A. et al. 2005. "Use of an antibiotic growth promoter and two herbal natural feed additives with and without exogenous enzymes in wheat-based broiler diets. *South African Journal of Animal Science*, 35.1: 61–72.

Sheridan, C., 2011. Gene therapy finds its niche. *Nature Biotechnology*, 29(2): 121–128. doi:10.1038/nbt.1769.

Shete, P., 2023. Biotechnological application in animal nutrition and feed technology: Mini review. *Acta Scientific Veterinary Sciences,* 5(5): 67–70.

World Health Organisation (OIE), 2010. *Definition of Animal Welfare, Glossary: Terrestrial Animal Health Code* (p. xiv). OIE.

5 Advancements of Plant and Animal Biotechnology

Sana Sheikh, Hariprasad Shetty, Jyothi Miranda, Sareen Sheikh, Renita Dsouza, and Shameena Karnad Abdul Rahiman

5.1 INTRODUCTION

Biotechnology is the technological application that utilizes biological entities or their derivatives for definite purpose. Biotechnology has caused a transformation in medical fields, agricultural sectors and various enterprises with sustainable aspirations. Biotechnology comprises molecular techniques to modify living organisms or create new processes or products which can have constructive influence on human and animal health, agricultural products and environment. It has ushered a new-fangled epoch in disease diagnosis, treatment and prevention. It has also facilitated tremendous advances in almost everything from groundbreaking cures that revolutionise medical treatments to sustainable agriculture practices that increase food output. Innovative approaches like gene editing, synthetic biology and personalised medicine are altering the healthcare landscape by providing targeted and specific therapies.

Plant and animal biotechnology research is critical to the production and conservation of plant and animal resources worldwide. According to the United Nations Environment Programme (UNEP) Convention on Biological Diversity, the word biotechnology is well defined as a set of techniques that uses biological systems, living organisms, or derivatives to create or modify products or processes for specific use (UNEP 1992).

One of the most important goals of nations in the arena of plant and animal biotechnology is to modify the standards of high-quality research output from universities and research institutes in the direction of national economic development (Moyo et al. 2011). This, however, has not been achieved as required in some nations due to various reasons, including a lack of entrepreneurial culture among researchers, which is mostly due to the high expense of securing foreign patents. Furthermore, technological difficulties, societal considerations, public perception of biotechnology, policy challenges and aspects linked to research aims and distribution of research products are also some of the hurdles to achieving this goal.

Biotechnology in health uses nature's strength and our understanding of biology to create novel solutions for disease prevention, diagnosis and treatment (Ma et al. 2017;

 DOI: 10.1201/9781003452805-5

Bao et al. 2017). It is very important that plant biotechnology research should focus on national needs if the technology is to be applied in a beneficial manner. A good collaborative effort between end users, industries and researchers could be helpful to scientists in applying biotechnological methods to meet the needs of countries (Moyo et al. 2011). Importance should be given to less commercial indigenous species of crops which are generally neglected, in addition to the major crops (Bornman et al. 2004). Such indigenous species could be developed as new crops with a large commercial potential. In funding biotechnology research projects, close attention should be given to the relevance of the proposed project to the national/regional needs (Moyo et al. 2011).

5.2 PLANT BIOTECHNOLOGY INNOVATIONS, OPPORTUNITIES AND ACHIEVEMENTS

Food bioengineering is a powerful and potentially beneficial technology. The potential benefits are feeding the world's hunger, improving agricultural production to make it cheaper and easier to grow crops and making possible new varieties of foods with desirable characteristics (e.g., improved taste, appearance or nutrition).

Plant biotechnology is also utilised for non-food applications in developed nations, in addition to the manufacturing of food, including paper, textile, biofuel, pharmaceutical and starch (Ricroch and Hénard-Damave 2016). Plant biotechnology has a wide range of applications (Figure 5.1). In such sectors, plants are modified to produce molecules with therapeutic uses, or with an improved biomass conversion efficacy or for manufacturing larger volumes of feedstocks for biofuels. A transgenic plant has a gene or genes that have been artificially added. Various plant breeding technologies are used, such as transgenesis producing proteins or RNA interference.

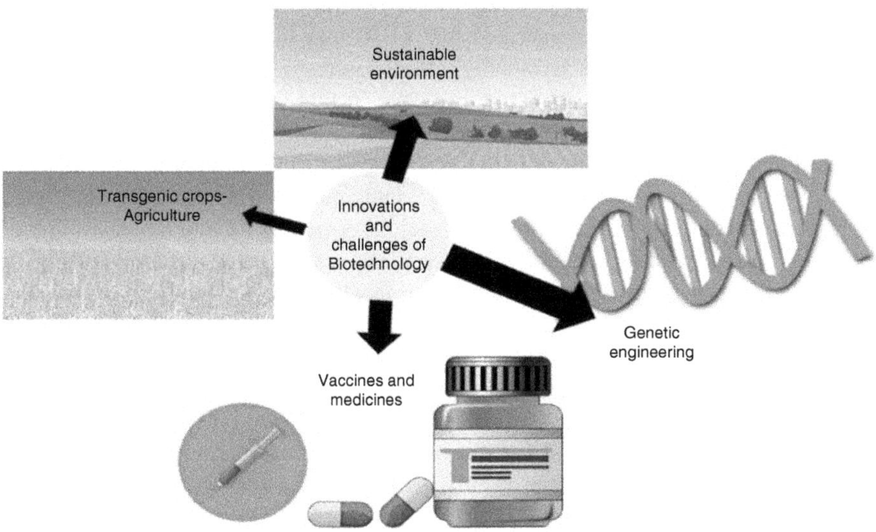

FIGURE 5.1 Broad applications of plant biotechnology.

Cisgenesis (transgenes isolated from a crossable donor plant) as well as intragenesis (transgenes originated from the same species or a crossable species), and null segregants are some of the examples of plant breeding technologies. Clustered regularly interspaced short palindromic repeats (CRISPR), oligonucleotide-directed mutagenesis, transcription activator-like effector nucleases and zinc finger nuclease (ZFN) are some of the next-generation precision gene editing tools developed in basic research (Ricroch and Hénard-Damave 2016).

5.2.1 Genome Editing in Biotechnology for Agriculture

The agricultural biotechnology industry is characterized by heavy investment in research and development, dynamic technical change and increasing concentration in both the output market and the ownership of intellectual properties that support ongoing innovation. Genome editing, which uses designed nucleases to modify specific regions of the genome, is undoubtedly a significant advancement in molecular plant breeding. Through base additions, deletions, gene replacement, or transgenic insertion, genome editing with designed nucleases is an effective method for producing beneficial novel phenotypes in crops (Wolt et al. 2016). According to Wolt et al. (2016), these strategies produce phenotypic variation in plants that is indistinguishable from that produced by natural means.

Indeed, once the function of a gene is known, it is possible to modify or transfect it. For example, herbicide-tolerance or insect-resistance gene, abiotic factors-tolerance gene such as drought or water stress, could be inserted in a plant of interest, conferring the desired trait to the plant. The second option is to knock out a gene which confers agronomic advantages to the plant. In 2005, the research group led by Voytas (Gabriel and Voytas 1997) in the United States initiated the use of ZFNs for genome editing in plant cells in tobacco protoplasts. They found out that gene insertion frequencies may be enhanced by up to 10% in comparison with random integration (Wright et al. 2005). Lloyd et al. (2005) demonstrated that ZFNs may be used to generate INDELs (Insertion and Deletions) in an artificially introduced restriction site in the *Arabidopsis thaliana* genome. In 2009, these techniques were further improved and aided to modify agronomic characters. Townsend et al. (2009) used ZFNs to induce SNPs via HDR in the SuRB loci in tobacco, establishing resistance to various herbicides. ZFN-based technologies have been then used to modify plant species, with a success in targeted editing ranged from 2% to 25% in plants such as *Arabidopsis* and soybean (Quetier 2016). New plant breeding technologies, especially site-directed genome editing, are workable substitutes to transgenic crop production which provide new opportunities for innovation (Wolt et al. 2016).

According to a literature study, the primary use of CRISPR systems in rice is to improve yield performance, biofortification, and biotic and abiotic stress tolerance (Ricroch et al. 2017). Since the first transgenic plant placed on the market in 1996, the "GMO" issue is predisposed to controversy. However, technology is constantly improving, new techniques appear rapidly as it is the case for genome editing with engineered nucleases (Wolt et al. 2016).

Besides, agriculture, research and seed industry also need these new molecular techniques. However, innovations and improvement depend on the size of the targeted market.

Conversely, a large scientific agreement is rapidly growing in favour of a product-based approach. This move is more realistic and would be a real revolution. This approach adopted by North American regulators involves regulating gene-edited crop varieties based on the trait instead of the method used to generate it (Wolt et al. 2016).

When genome editing is used for plant breeding, the outcome of genetic modifications can be assessed, but it is not always possible to know which technique to be used. Based on current legislation, taking into account scientific advices and the JRC's advice in 2012, edited plants should not be considered as GMOs when engineered nucleases are used for targeted mutagenesis, but should be considered as GMOs when engineered nucleases are used for gene insertion. The United Kingdom criticized EU policy for being in favour of adopting a product-based approach. Sweden did not wait for EU agreement to conduct field trials of edited plant and authorized trials which used engineered nucleases to carry out targeted mutagenesis. Finally, France and Germany gave advices which are in line with the idea that plants which have no foreign DNA, and point mutations should not be considered as GMOs (Pfeiffer et al. 2018).

The European Commission decision is still under discussion. Since seed industry and plant biotech research are important for the EU, the categorization of gene-edited plants as non-GM would be a serious advantage for the EU. If European Commission classifies gene-edited crops as GMOs, the EU should import gene-edited crops or food produced from or using gene-edited crops (Pfeiffer et al. 2018).

5.2.2 Crops with Genetic Modifications

Cultivation of genetically engineered (or "genetically modified" as legally termed in some countries; GM) crops has increased markedly since the first introduction of the transgenic technology in 1996, representing 215 million of hectares (about 18.5% of all cropland worldwide) (www.isaaa.org 2022). Of the 14 genetically modified traits approved for commercial production, the utmost extensively planted genetically modified crops are those expressing insecticidal toxins targeted at specific insect pests or those tolerant to one or more herbicides. Insect-resistant crops or so-called Bt maize, cotton and soybean have been genetically modified to express one or more insecticidal proteins from the natural occurring soil bacterium *Bacillus thuringiensis* (Bt). These Cry (δ-endotoxins found in crystals formed at the sporal stage) or VIP (vegetative insecticidal proteins) toxins bind to specific epithelial receptors in the gut wall of insect larvae, causing feeding cessation, cell lysis and pore formation that eventually result in gut paralysis and death. The specific binding of the particular Bt protein exerts a selective toxic effect on a specific group of target insects or taxonomic order, such as *Lepidoptera* (moths) or *Coleoptera* (beetles) in the case of current Bt crops. Transgenic proteins expressed in GM plants for controlling insect pests include protease inhibitors (PIs). When ingested by insects, specific proteases can inhibit the digestive proteolytic enzymes in the insect gut, resulting in starvation and eventual death. Nevertheless, nothing have been commercially planted to date with the exception of a PI trait in China (Ricroch et al. 2018).

Herbicide-tolerant crops have been genetically engineered with plant/bacterial genes encoding specific enzymes that either detoxify the herbicide or provide a metabolic backup pathway to tolerate the herbicide (Ricroch et al. 2018). Feeding the world can be done through the existing scientific innovations including gene editing, but achieving it depends on regulations.

5.2.3 Plant Biotechnology and Biopharmaceuticals

Around the world, the COVID-19 epidemic has caused a serious public health catastrophe. Using a quick, scalable technique called transient/stable expression in plants, some plant biotechnology researchers collaborating with commercial companies for emergency diagnostics and treatments sought to satisfy the growing demand for SARS-CoV-2 protein antigen and antibody (Shohag et al. 2021). Plant biotechnology with transient/stable expression provides a quick response to this dilemma by producing low-cost diagnostics, antiviral medicines, immunotherapy and vaccines. Transient/stable expression technique for the production of plant-based biopharmaceuticals is already commercially available.

Shortly after the first successful expression of a recombinant antibody in a plant, plants were first used as a biopharmaceutical platform for the synthesis of therapeutic proteins and diagnostic reagents (Hiatt et al. 1989). Plants offer huge advantages compared with platforms based on traditional mammalian cell cultures. When compared to conventional mammalian cell cultures, plants have many advantages. Within 2 months of obtaining the correct RNA sequence, recombinant protein engineering and production in transgenic plants can be achieved by a transient expression system mediated by agroinfiltration and/or viral vectors (Gleba et al. 2013; Shoji et al. 2012).

In light of the present pandemic issue, plants may be the platform that may be utilised to produce diagnostics and treatments on a large scale in a matter of weeks. They can also be swiftly ramped up to meet sudden and unexpected demands by using transient expression systems. Plant-based proteins for diagnostics and treatments are the focus of several biopharmaceutical companies, such as Protalix (Carmiel, Israel), Bayer (Halle, Germany), Planet Biotechnology (Hay Ward, CA, USA), Ventria (Fort Collins, CO, USA), Fraunhofer CMB (Newark, DE, USA), Medicago Inc. (Quebec, QC, Canada), Agrenvec (Madrid, Spain), Diamante (Verona, Italy), ORF Genetics (Kópavogur, Iceland), Ventria Bioscience/Invitria (Fort Collins, CO, USA) and Mapp Biotherapeutics (San Diego, CA, USA) (Shohag et al. 2021).

Flavonoids, terpenes, coumarins, lignans, alkaloids, phenolic acids, proteins, and other plant extracts, isolated bioactive compounds and essential oils have all demonstrated a critical role as antiviral medications (Daglia 2012). The emergence of some viral resistance has, however, led to a growing interest in plant-based protein manufacturing as a potential antiviral medication, as many conventional drugs have been proven to be less efficient against viral infection (Irwin et al. 2016).

In this instance, passive immunotherapy using plants to manufacture antibodies could be employed. During the Zaire Ebolavirus outbreak in West Africa, Mapp Biopharmaceutical (San Diego, CA, USA) successfully demonstrated this process for the first time. The LeafBio commercial division of Mapp Biopharmaceutical created the ZMapp cocktail of three neutralising antibodies (Hiatt et al. 2014), which was authorised because of its life-saving properties and the absence of any substitutes (Na et al. 2015; Qiu et al. 2014).

Because up to 10 mg of antibodies is required per infected patient, mass-producing transgenic plants would be a cost-effective way to make such a product. Medicago Inc.

(Québec, Canada) recently announced their desire to employ their plant-based platform for the rapid manufacture of antibodies against SARS-CoV-2 in partnership with Laval University's Infectious Disease Research Centre. These SARS-CoV-2 antibodies have the potential to be employed in the treatment of severely infected hospitalised patients. Transient/stable expression in plants is a faster, more cost-effective, scalable and flexible technology than traditional microbe-, insect- or mammalian cell-based platforms because there is no need to establish stable culture cell lines or expensive culture media, nor is there any additional set-up for scaled-up production except for the cultivation of more plants (Shohag et al. 2021).

Furthermore, the creation of stable transformation is required in order to make edible vaccines, which may find a place in the post-pandemic phase. Crop plants may be grown in a variety of settings; hence, biopharmaceuticals could be manufactured utilising existing established agricultural infrastructures and the same distribution networks that exist for the provision of food and cereal seeds, eliminating the requirement for a refrigerated supply chain. Plant biotechnology has the potential not only to combat the recent crisis, but also to develop a perfect model that allows for a swift and planned reaction to any future crises (Shohag et al. 2021).

5.3 ANIMAL BIOTECHNOLOGY INNOVATIONS, OPPORTUNITIES AND ACHIEVEMENTS

Animal biotechnology is a branch of biotechnology which employs various techniques to genetically engineer (i.e., modify) the genome of animals so that it improves their suitability for agriculture, industrial or pharmaceutical applications. Animal biotechnology innovation encompasses a wide range of applications, from genetic engineering to cloning and breeding techniques.

Biotechnology increases the nutritional value of food we consume like milk, meat and eggs. It also plays a vital role in detection, treatment and prevention of diseases. It allows farmers to have healthier offsprings of various breeds and therefore produce safer and higher quality foods consistently. Animals play a vital role in the growth of biotechnology.

Predominantly innovations and achievements are in the following areas as mentioned in Figure 5.2.

5.3.1 Animal Biotechnology to Advance Human Health

Animals have been used for years to produce medicines for humans. Animal-made pharmaceuticals transform transgenic animals into "factories" to produce therapeutic proteins in their milk, eggs and blood, which can be used in the development of biopharmaceuticals.

Biopharming involves using transgenic animals to produce (or "farm") pharmaceutical products for therapeutic use.

- This involves the insertion of target genes into hosts (animals) that would not normally express those genes.

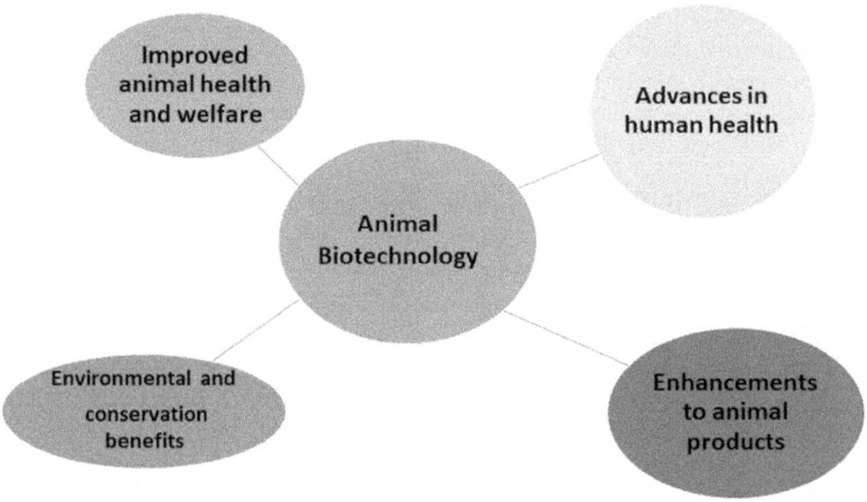

FIGURE 5.2 Innovations and achievements in animal biotechnology.

- The desired compound can potentially be expressed in a form that is routinely harvested (e.g., milk, eggs, etc.).

  Various examples of biopharming, including transgenic sheep that produce human α-1-antitrypsin in their milk (individuals deficient in this enzyme develop emphysema) are reported.

In addition, biotechnology can be used to produce human-compatible transplant organs, tissues and cells in pigs that can be vital to enhancing human health.

5.3.2 Biotechnology to Improve Animal Health

Livestock have been improved for years, through enhanced animal husbandry practices and modern technologies. These include methods such as artificial insemination, embryo transfer, in vitro fertilization, genetic mapping and cloning. Biotechnology helps farmers to enhance breeding, resulting in healthier herds. Also, treatments have been developed by the animal health industry that can prevent and treat disease. Farmers can treat food-borne pathogens and various animal diseases by using new vaccines and diagnostic tests.

5.3.3 Biotechnology to Develop More Nutritious Food

Food quality may be improved by introducing desirable traits into farm livestock and poultry. In the future, meat, milk and egg products from animals can be nutritionally enriched with the use of biotechnology. The new wave of technological changes has determined intense improvements in various sectors from production of drugs, vitamins, steroids, interferon, to desirable nutritious food via genetic engineering.

5.3.4 Conservation of Environment and Animals

Biotechnology can help create GM organisms who are environmentally friendly, and in conserving endangered species. Feeds of farm animals have been improved by employing biotechnology to reduce animal wastes, minimizing the impact on the environment. Genetics of endangered species can be preserved by using reproductive cloning techniques. Genetic studies of endangered animals can also result in increased genetic diversity with healthier populations of species.

Animal biotechnology can play a vital role in conserving endangered species through techniques like cloning or assisted reproduction. The conservation of endangered wildlife species has seen the emergence of powerful strategies for in situ and ex situ conservation programs (Andrabi and Maxwell 2007). It is a component of aesthetic value of biodiversity.

5.3.5 Applications in a Nutshell

Animal biotechnology comprises of variety of animals like livestock, poultry, fish, insects and laboratory animals.

Applications of animal biotechnology are in the following sectors:

5.3.5.1 Animal Genomics

Genomics makes up the complete genetic makeup of an animal. Knowledge of genomics helps us to understand the basis for disease resistance, disease susceptibility, weight gain and determinants of nutritional value.

5.3.5.2 Animal Cloning

Livestock breeders can create an exact genetic copy of an existing animal using the technique of somatic cell nuclear transfer. Cloning is a form of sophisticated assisted reproduction.

5.3.5.3 Genetic Engineering of Animals

A **transgenic animal** is one which has genetic material from another species incorporated into its DNA. This technology allows scientists to precisely transfer beneficial genes from one species to another. Genetic modification can help develop animals resistant to **specific diseases**, reducing the need for antibiotics in agriculture. Researchers and breeders have traditionally prioritised genetic improvement in farm animals. Several biotechnologies have been created and successfully implemented in animal production to improve efficiency, hence influencing genetic advancement. These technologies are critical not just for meeting worldwide demand for animal products, but also for preserving biodiversity and genetic variation in threatened species. They boost fertility, product quality, growth and nutrient efficiency. Artificial insemination, embryo transfer, in vitro fertilisation, cloning and genome editing are all reproductive biotechnologies that have a substantial impact on germplasm spread and selection intensity. They can, however, indirectly limit genetic variance, which is essential for genetic change. Novel biotechnologies like CRISPR/Cas9 genome editing and transgenesis hold potential for future advancements in animal breeding (Das et al. 2022).

Newer techniques from domestic species, like intracytoplasmic sperm injection and nuclear transfer, are promising for the propagation of endangered species (Comizzoli et al. 2000). The development of livestock breeds involves genetic changes influenced by the environment and human intervention. Initially, the diversity of farm animal breeds increased due to favourable conditions and human demands. This diversity is maintained through genetic and biotechnological methods like assisted reproduction and cryopreservation. Efforts are focused on improving research efficiency to create biological reservoirs for various applications in biotechnology, biomedicine and the pharmaceutical industry.

5.4 ETHICAL INTERROGATIONS IN BIOTECHNOLOGY

The ethical issues related to genetic modification in research and its future clinical applications are examined (Munsie and Gyngell 2018). The developments in genome modification technologies, particularly the CRISPR/Cas9 system, relates to ethical examinations. It covers the interrogations about gene editing in human embryos and the concerns about altering the germline. Various international groups and reports have highlighted the importance of ethical considerations in genome editing, distinguishing between basic research and clinical applications. Various people have interrogated ethical concerns towards environmental impact and bioremediation. Numerous groups have discussed the use of GMOs in bioremediation to remove pollutants along with relevant laws related to GMOs in environmental clean-up.

Improving research's translational relevance, as discussed by Landi et al. (2021), could avert pointless studies and raise the success rate of clinical trials involving humans. The ethical obligations to science, animals and society are all in line with the need to guarantee the reproducibility and translation of animal research. Moving away from animal testing, alternatives such as in vitro procedures, volunteer studies and computer modelling gain prominence due to ethical, scientific and legislative concerns (Langley et al. 2007). Despite limited progress in substituting animal testing, methodical assessments on the applicability of animal data in therapeutic contexts should encourage the investigation of innovative non-animal methods. Increasing popular financing could help support ongoing improvements in this field. While lower animal models, like invertebrates and lower vertebrates, are appealing for toxicity evaluation due to their cost-effectiveness and simplicity (Khabib et al. 2022), standardized methodologies are crucial for repeatability. Correct experimental data depends on well-designed husbandry settings. Research results from animal trials heavily influence clinical trial tactics, necessitating efforts to reduce bias and random errors (Asokan et al. 2012). To handle common dangers, especially zoonoses, the One Health concept supports collaboration between human and animal health practitioners. Standardizing research methodologies, conducting power analyses, prospectively registering investigations and performing systematic reviews are crucial for ensuring comparability between animal models and clinical trials. Integrating results from disparate studies improves the efficacy and accuracy of animal research, increasing its relevance to human health (Singh et al. 2016). Critical to this is identifying bias and removing confounding variables during statistical analysis in animal study designs. Addressing the volatility of effects is also necessary.

To increase the accuracy of experiments in biomedical research, methods must be improved and training in laboratory animal science must be given. This emphasizes the significance of paying attention to details frequently missed while formulating hypotheses and designing experiments.

5.5 CHALLENGES

Meeting global food security concerns is critical in developing countries. Plant biotechnology is a tool for producing high yields in the face of biotic and abiotic stressors. Food security is not just a matter of quantity, but also of the quality of agricultural and food items provided to those in need. Plants having nutritional properties, such as biofortification in vitamins and nutrients, are thus the new norm in plant biotechnology (Ricroch and Hénard-Damav 2016).

Despite the fact that cell cultures for many plants have been produced, commercial success is restricted due to various technological constraints such as low productivity and bioprocess engineering. The bioreactor technology remains relatively expensive for underdeveloped nations, impeding the scaling-up process for industrial secondary metabolite synthesis. Understanding the principles of biosynthetic pathways and the regulatory mechanisms that control plant secondary metabolism aids in the exact regulation of biochemical processes for greater productivity (Moyo et al. 2011).

Inadequate infrastructure for scaling up bioreactor technology, a lack of competent human resources and the high cost of technological development and transfer are only a few of the significant challenges that developing nations confront. This problem is exacerbated by a lack of job prospects for graduates in the academic and industrial sectors, as well as a general lack of interest in this profession. Some of the measures for human resource growth and retention include an aggressive international recruitment programme and the construction of competitive salary packages (Moyo et al. 2011).

Plant biotechnology companies are distinguished by the need for multiple rounds of finance before becoming self-sustaining. In underdeveloped nations, a lack of appropriate start-up funding or investment capital is a limiting factor for the development and expansion of the plant biotechnology business. However, just a few underdeveloped countries have undertaken start-up funding. This problem is described as a serious impediment to progress in the field of plant biotechnology. Biotechnology firms in poor nations receive less support than those in the United States and Europe. This could explain why there are not many plant biotechnology business sectors or firms in developing nations where research can be turned into monetary value. As a result, a more robust finance system would undoubtedly aid in overcoming economic issues.

A review of the literature reveals that several research techniques for medicinal and/or decorative plants have been created. However, the majority of these protocols end up in academic publications, and the outputs frequently do not reach end-users who have little or no access to these published pieces. The increased volume of research publications published has also resulted in research that lacks market or commercial value. The desire to publish research articles in high-impact scientific journals has thus become the primary driving force in scientific communities, with little or no regard for commercialization or market value.

5.5.1 Challenges—Plant Biotechnology

Pollen or genetic drift from transgenic plants to other plant species, resistance issues like superweeds, resistance to insects and diseases, unknown long-term consequences of human and ecological health, possible allergic reactions to new products, the profit-making focus of the biotechnology industries, corporate control of food systems, enhancing an organism's ability to become an invasive species, the potential for new diseases to emerge, gene flow from herbicide-resistant crops to wild species and loss of biodiversity are some of the challenges in the field of plant biotechnology.

There are concerns expressed about the potential impacts of releasing GMOs into the environment and the study of the possible negative side effects to monarch butterflies from Bt corn pollen.

5.5.2 Challenges in Detecting and Quantifying GM Animals

An effective enforcement system for detecting GM animals necessitates accurate sampling, reliable detection and quantification methods and access to reference materials. While detecting GM animals presents no significant challenges, quantification becomes more complex when dealing with single-copy nuclear endogenes and mosaic or non-homozygous animals. The importance of these challenges is case-specific and requires a thorough examination with complete sequence information (Lievens et al., 2015).

5.5.3 Challenges in Animal Genetic Modification and Cloning

Ethical Concerns: Animal ethics are a key concern, particularly in genetic manipulation and cloning. Creatures' welfare and rights must be properly regarded. The intersection of genetics and ethics in the context of genetic alteration and genome editing should be investigated. While these technologies have the potential to be beneficial, there are complications and ethical issues to consider. Genome editing, which is sometimes promoted as a simple repair, is hampered by a lack of genetic understanding. Honest and transparent communication, and ethical exchanges with stakeholders are essential. Discussions and judgements about animal breeding in the context of genetic modification and genome editing should include ethical issues. The authors advocate for the establishment of ethical advisory committees inside breeding organisations, as well as the revision of industry standards of conduct to accommodate new technologies. These debates are critical (Das et al. 2022).

Welfare of Agricultural Animals: Genetic modification technology can be employed to enhance the welfare of agricultural animals. Here, the potential of using genetic modification technology to enhance the welfare of agricultural animals has been discussed. Genetically modifying animals for improved welfare can be a preferable option compared to the current conventional practices. However, it is acknowledged that alternative approaches, such as plant-based diets or *in vitro* meat, may have ethical advantages. Despite these alternatives, the chapter argues for the continued pursuit of welfare improvements through genetic modification due to compelling reasons. Assisted reproductive techniques are crucial for propagating small populations of both wild endangered species and domestic breeds. They prevent inbreeding depression by producing offspring from selected individuals. However, adapting methods to various species and constraints is challenging. Genome resource banking is emerging, and artificial insemination is commonly used (Shriver and McConnachie 2018).

5.6 CONCLUSION

In this book chapter, we have delved into the vast landscape of animal biotechnology, exploring its multifaceted applications, from genetic engineering to breeding techniques. We have examined the numerous opportunities it presents, such as improved livestock production, its potential for enhancing human health through transgenic animals and its role in conservation, disease resistance, and welfare improvement of agricultural animals. However, it is important to recognize that alongside these opportunities, there are also considerable challenges in the field of animal biotechnology. Ethical concerns regarding the treatment of animals in genetic modification and cloning must be carefully addressed. As technology advances, it becomes increasingly crucial to integrate ethical considerations into discussions and decisions related to animal breeding. We have also explored the environmental and regulatory aspects of animal biotechnology, highlighting the importance of responsible practices and the need for an ongoing discussion about public perception and acceptance. As we move forward in this ever-evolving field, it is essential to balance scientific progress with ethical responsibilities, environmental considerations and public awareness. The responsible and ethical use of animal biotechnology can lead to profound benefits in agriculture, human health and conservation, while mitigating potential risks and concerns.

In conclusion, animal biotechnology holds tremendous promise, but its success depends on a harmonious synergy between science, ethics and society. As researchers, policymakers and industry professionals continue to navigate this complex landscape, they must keep these considerations at the forefront of their efforts to ensure a brighter and more sustainable future for all living beings.

Over time, the increasing pace of technological innovation ensures that newer and better technologies continue to be developed. Often, it is the regulators that must play catch-up with scientific and technological developments. Advancements in technology are happening at a much faster pace today than almost any other time in human history. Increased global literacy, globalisation, rapid growth in information technology services, automation and artificial intelligence are all contributing to the accelerated pace of innovation.

To secure sustainable, ethically responsible and socially legitimate use of biotechnology, regulation must be in accordance with public values. The regulation of biotechnological innovation should be knowledge-based while at the same time reflecting our values. To achieve this, there is a need for empirical research into the outcome of diverse forms of engagement, as well as normative research into the ethical values and norms at stake in biotechnology innovation and regulation.

REFERENCES

Andrabi, S.M.H. and Maxwell, W.M.C., 2007. A review on reproductive biotechnologies for conservation of endangered mammalian species. *Animal Reproduction Science*, *99*(3–4), pp. 223–243.

Asokan, G.V., Fedorowicz, Z., Tharyan, P. and Vanitha, A., 2012. One health: perspectives on ethical issues and evidence from animal experiments. *Eastern Mediterranean Health Journal*, *18*(11), pp. 1170–1173.

Bao, W., Jiang, Z. and Huang, D.S., 2017. Novel human microbe-disease association prediction using network consistency projection. *BMC Bioinformatics*, *18*, pp. 173–181.

Bornman, C.H., Grace, O.M. and Van Staden, J., 2004. Sustainable biotechnology for sub-Saharan Africa: can it be implemented and maintained? *South African Journal of Botany*, *70*(1), pp. 1–11.

Comizzoli, P., Mermillod, P. and Mauget, R., 2000. Reproductive biotechnologies for endangered mammalian species. *Reproduction Nutrition Development*, *40*(5), pp. 493–504.

Daglia, M., 2012. Polyphenols as antimicrobial agents. *Current Opinion in Biotechnology*, *23*(2), pp. 174–181.

Das, D.N., Paul, D. and Mondal, S., 2022. Role of biotechnology on animal breeding and genetic improvement. In: S. Mondal and R. L. Singh (Eds.), *Emerging Issues in Climate Smart Livestock Production* (pp. 317–337). Academic Press.

Gabriel, A. and Voytas, D., 1997. DNA on the move. *Trends in Genetics*, *13*(7), pp. 258–259.

Gleba, Y.Y., Tusé, D. and Giritch, A., 2013. Plant viral vectors for delivery by *Agrobacterium*. In K. E. Palmer and Y. Gleba (Eds.), *Plant Viral Vectors* (pp. 155–192). Springer.

Hiatt, A., Caffferkey, R. and Bowdish, K., 1989. Production of antibodies in transgenic plants. *Nature*, *342*(6245), pp. 76–78.

Hiatt, A., Pauly, M., Whaley, K., Qiu, X., Kobinger, G. and Zeitlin, L., 2014. The emergence of antibody therapies for Ebola. *Human Antibodies*, *23*(3–4), pp. 49–56.

Irwin, K.K., Renzette, N., Kowalik, T.F. and Jensen, J.D., 2016. Antiviral drug resistance as an adaptive process. *Virus Evolution*, *2*(1), p. vew014.

Khabib, M.N.H., Sivasanku, Y., Lee, H.B., Kumar, S. and Kue, C.S., 2022. Alternative animal models in predictive toxicology. *Toxicology*, *465*, p. 153053.

Landi, M., Everitt, J. and Berridge, B., 2021. Bioethical, reproducibility, and translational challenges of animal models. *ILAR Journal*, *62*(1–2), pp. 60–65.

Langley, G., Evans, T., Holgate, S.T. and Jones, A., 2007. Replacing animal experiments: choices, chances and challenges. *Bioessays*, *29*(9), pp. 918–926.

Lievens, A., Petrillo, M., Querci, M. and Patak, A., 2015. Genetically modified animals: options and issues for traceability and enforcement. *Trends in Food Science & Technology*, *44*(2), pp. 159–176.

Lloyd, A., Plaisier, C.L., Carroll, D. and Drews, G.N., 2005. Targeted mutagenesis using zinc-finger nucleases in Arabidopsis. *Proceedings of the National Academy of Sciences*, *102*(6), pp. 2232–2237.

Ma, W., Zhang, L., Zeng, P., Huang, C., Li, J., Geng, B., Yang, J., Kong, W., Zhou, X. and Cui, Q., 2017. An analysis of human microbe-disease associations. *Briefings in Bioinformatics*, *18*(1), pp. 85–97.

Moyo, M., Bairu, M.W., Amoo, S.O. and Van Staden, J., 2011. Plant biotechnology in South Africa: micropropagation research endeavours, prospects and challenges. *South African Journal of Botany*, *77*(4), pp. 996–1011.

Munsie, M. and Gyngell, C., 2018. Ethical issues in genetic modification and why application matters. *Current Opinion in Genetics & Development*, *52*, pp. 7–12.

Na, W., Park, N., Yeom, M. and Song, D., 2015. Ebola outbreak in Western Africa 2014: what is going on with Ebola virus? *Clinical and Experimental Vaccine Research*, *4*(1), pp. 17–22.

Pfeiffer, M., Quétier, F. and Ricroch, A., 2018. Genome editing in agricultural biotechnology. In: J.-P. Jacquot and P. Gadal (Eds.), *Advances in Botanical Research* (vol. 86, pp. 245–286). Academic Press.

Qiu, X., Wong, G., Audet, J., Bello, A., Fernando, L., Alimonti, J.B., Fausther-Bovendo, H., Wei, H., Aviles, J., Hiatt, E. and Johnson, A., 2014. Reversion of advanced Ebola virus disease in nonhuman primates with ZMapp. *Nature*, *514*(7520), pp. 47–53.

Quetier, F., 2016. The CRISPR-Cas9 technology: closer to the ultimate toolkit for targeted genome editing. *Plant Science*, *242*, pp. 65–76.

Ricroch, A., Akkoyunlu, S., Martin-Laffon, J. and Kuntz, M., 2018. Assessing the environmental safety of transgenic plants: honey bees as a case study. In: J.-C. Kader and M. Delseny (Eds.), *Advances in Botanical Research* (vol. 86, pp. 111–167). Academic Press.

Ricroch, A.E. and Hénard-Damave, M.C. Next biotech plants: new traits, crops, developers and technologies for addressing global challenges. *Critical Reviews in Biotechnology*, *36*(4), 675–90. doi:10.3109/07388551.2015.1004521.

Ricroch, A., Clairand, P. and Harwood, W., 2017. Use of CRISPR systems in plant genome editing: toward new opportunities in agriculture. *Emerging Topics in Life Sciences*, *1*(2), pp. 169–182.

Shohag, M.J.I., Khan, F.Z., Tang, L., Wei, Y., He, Z. and Yang, X., 2021. COVID-19 crisis: how can plant biotechnology help? *Plants*, *10*(2), p. 352.

Singh, V.P., Pratap, K., Sinha, J., Desiraju, K., Bahal, D. and Kukreti, R., 2016. Critical evaluation of challenges and future use of animals in experimentation for biomedical research. *International Journal of Immunopathology and Pharmacology*, *29*(4), pp. 551–561.

Shriver, A. and McConnachie, E., 2018. Genetically modifying livestock for improved welfare: a path forward. *Journal of Agricultural and Environmental Ethics*, *31*, pp. 161–180.

Townsend, J.A., Wright, D.A., Winfrey, R.J., Fu, F., Maeder, M.L., Joung, J.K. and Voytas, D.F., 2009. High-frequency modification of plant genes using engineered zinc-finger nucleases. *Nature*, *459*(7245), pp. 442–445.

UNEP, 1992. *UNEP Convention on Biological Diversity, Article 2*. United Nations Environment Programme. https://www.cbd.int.

Wolt, J.D., Wang, K. and Yang, B., 2016. The regulatory status of genome-edited crops. *Plant Biotechnology Journal*, *14*(2), pp. 510–518.

Wright, D.A., Townsend, J.A., Winfrey Jr, R.J., Irwin, P.A., Rajagopal, J., Lonosky, P.M., Hall, B.D., Jondle, M.D. and Voytas, D.F., 2005. High-frequency homologous recombination in plants mediated by zinc-finger nucleases. *The Plant Journal*, *44*(4), pp. 693–705.

6 Plant and Animal Biotechnology Innovation Action Plan

Main Opportunities and Challenges

Kapil Paul

6.1 INTRODUCTION

The use of living things or their by-products for the betterment of humans is known as biotechnology. It has revolutionized all branches of biology and flourished since primordial times (Bhatia 2015). The benefits of biotechnology have touched many practical domains, ranging from agriculture to health of humans as well as animals (AL-Eitan and Alnemri 2022). Biotechnology enables us to create a genetically modified plant by taking genetic information from another plant, altering it in the laboratory, and then putting it into a plant to change its certain properties. Using ancient knowledge, classical biotechnology makes extensive use of technologies tailored to industrial manufacturing. Many approaches developed in classical biotechnology are still widely employed today. Initiatives have been made to improve understanding of microscopy techniques, biochemistry, and genetic modification in organisms through associated disciplines and technology. This introduces a new period known as the modern genetics' era (Narayanaswamy 1994; Mishra 2009). This modern era of biotechnology has enabled us to perform numerous tasks as discussed ahead (Table 6.1).

6.2 OPPORTUNITIES IN PLANT BIOTECHNOLOGY INNOVATION ACTION

The use of living things or their by-products for the betterment of human as well as environmental health is known as biotechnology. Biotechnology utilizes the manufacturing ability of cells and biological substances like proteins and DNA to create useful products (Bhatia 2015). Plant biotechnology and plant science refer to biological processes that use plant cells to develop useful plants and products (Bhojwani and Razdan 1996; Jha and Ghosh 2005). Plant biotechnologists have great opportunities for advancement in agriculture and the development of new variants of crops with better quality, high yield, disease resistance, etc. Biotechnology brings about the following opportunities to meet the needs of present as well as future (Figure 6.1).

DOI: 10.1201/9781003452805-6

TABLE 6.1
Uses of Biotechnology in Plant and Animal Research

Aspect	Plant Biotechnology	Animal Biotechnology
Increased yield	Research and development of high-yield, climate-resilient, and pest-resistant crops	Improvement in livestock productivity (milk and meat yield)
Stress tolerance	Development of crops tolerant to salinity, temperature, and drought extremes	Development of vaccines and other healthcare products
Enhancement in nutritional value	Biofortification (products having enhanced nutritional values)	Increasing nutritional values of milk, egg, and meat
Pharmaceutical production	Production of plants-based pharmaceutical products	Use of animal products like bones, milk, and eggs to produce pharmaceutical products
Precision breeding	Use of gene editing tools and CRISPR for targeted gene improvement	Gene editing to treat genetic disorders
Environmental benefits	Reducing the use of fertilizers and pesticides	Better feed efficiency results in lower greenhouse gas emissions
Conservation	Seed banks and tissue culture for endangered plant species	Cryopreservation and cloning for endangered animal species

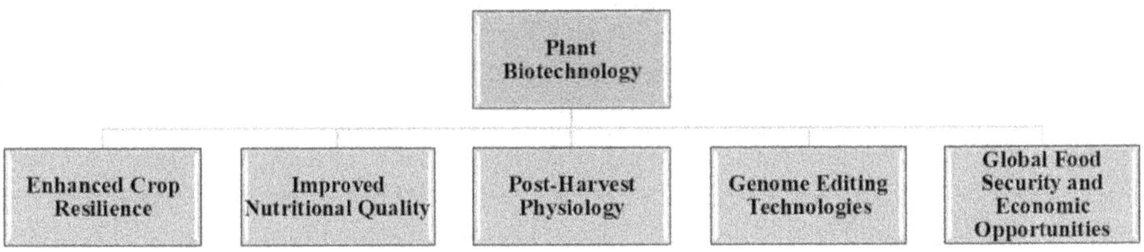

FIGURE 6.1 Opportunities in plant biotechnology.

6.2.1 Enhanced Crop Resilience

Crop plants encounter several obstacles, including variable weather patterns, developing pests, and illnesses. Improving crop plant resistance has become increasingly important for ensuring food security and sustainable farming practices (Chauhan et al. 2023). The ability of plants to withstand shocks and recover from them, leading to steady yields and overall health, is known as crop resilience. As global agriculture navigates the implications of climate change, evolving pest challenges, and the need for sustainable practices, agronomists are studying and developing techniques to improve crop resilience (Chauhan et al. 2023; Abraham et al. 2014). So, several measures are required for the enhancement of crop resilience, depending upon the type of stressor. Healthy soil is the fundamental need for growing crops. Providing healthy soil with good nutritive values is very important for crops. Therefore, cover cropping is one of the best methods to maintain soil health along with the suppression of weeds and prevention of soil erosion (Banga and Kang 2014). Adding organic compounds to soil, a strategy related to management of soil, is essential for improving nutrient cycling and microbial diversity. Organic matter's effects on soil structure, water retention, and nutrient availability highlight how crucial it is to create resilient soils that support robust crop growth. In pattern rainfall and water-scarce areas, availability of water is a big challenge for agriculture. To ensure the optimum availability of water to crops, rainwater harvesting and precision irrigation techniques such as sprinkling and drip systems should be used to minimize the effect of water stress on resilience of crop (Abraham et al. 2014). In agriculture, another tedious task that does not sacrifice crop quality or consumer health is pest management. Hence, disease-resistant varieties of crops should be developed using interbreeding and genetic engineering technologies (Wu, Shah, and Ma 2022). Biotechnological technologies, such as gene editing and modification, provide chances to create better crops. We highlight the potential of these techniques in generating resilient crops that can grow under difficult environmental conditions (Roberts and Mattoo 2018). Precision agriculture is likely to skyrocket, as technology advances. Emerging technologies, such as artificial intelligence, advanced robotics, and machine learning, are expected to perform critical roles in improving the resource utilization, decision-making, and crop monitoring. Incorporating these technologies into farming operations will not only increase efficiency, but will also help to construct resilient and adaptable agricultural systems (Usigbe et al. 2024; Liebman and Schulte 2015; Verhulst et al. 2011; Liu et al. 2022). A major abiotic stressor that affects plant growth, agricultural productivity, and development is salt, especially in regions with high saline soil. Growing crops that can withstand salt has become essential for maintaining food security as a result of climate change-induced soil salinization. Thus, improvements in plant genetics allow to use next-generation sequencing and transcription factors (TFs), which are essential for improving crop tolerance to salt stress through homeostasis, osmotic adjustment, and stress signaling pathways controlled by key genes that contribute to plant resilience in high salinity environments. Specific TF families, including DREB, NAC (NAM, ATAF, and CUC), and WRKY, play important roles in activating salt-responsive gene networks (Singh et al. 2024).

6.2.2 Improved Nutritional Quality

Around 870 million people are believed to suffer chronic hunger, malnutrition, and micronutrient deficiencies, a major global concern, particularly in developing countries (FAO, WFP, and IFAD 2012; Uncu, Doganlar, and Frary 2013). Biotechnology has played an important role in enhancing food and nutrition around the world. With an increasing population and changing climatic circumstances, traditional food production methods are no longer adequate to supply the growing demand for food. Biotechnology provides creative ideas for improving crop yields, increasing nutritional value, and addressing food security challenges (Farre et al. 2012). One essential biotechnology option for food and nutrition is genetic engineering. The process of altering an organism's genetic makeup to introduce desired features or attributes is known as genetic engineering. Genetically engineered crops that are resistant to pests, diseases, and environmental difficulties have been produced using this technique. These crops have the potential to boost agricultural output while also improving food security by providing a consistent and reliable food supply. Additionally, biotechnological techniques have been used to increase the nutritional value of crops. For example, biofortification is the process of genetically altering crops to increase their nutritional value. This technology has been used to create crops rich in important minerals and vitamins like Zn, Fe, and Retinol. In underdeveloped nations, biofortified crops have the potential to improve community health and well-being while also preventing malnutrition. Beyond rudimentary nutrition, functional foods offer extra health benefits including improving gastrointestinal health or reducing the threat of chronic diseases. Biotechnological technologies have been utilized to develop functional foods that address specific health issues like cardiovascular disease, diabetes, and obesity. Similarly, biotechnology has been employed to create nutritional supplements that contain important nutrients and promote general health and well-being (Kishore and Shewmaker 1999). Biopharmaceutical development is an important part of biotechnological solutions for food and nutrition. Biopharmaceuticals are a class of medications and cures developed by biotechnological technologies (cell culture and genetic engineering). Numerous medical conditions, such as diabetes, autoimmune diseases, and cancer, may be treated by these medications. Modern medicine has been revolutionized by biopharmaceuticals, which also hold promise for raising the standard of living for millions of people globally. Biotechnology has brought innovative and sustainable solutions to change the nutrition industry and has been utilized to improve food security, boost nutritional value, and promote environmental sustainability. Examples include genetic engineering, biofortification, food preservation, functional foods, and biopharmaceuticals. As the world's population grows, biotechnology solutions are essential to guaranteeing a safe, wholesome, and sustainable food supply for future generations. To meet the evolving needs of our food system, policymakers, researchers, and stakeholders must continue to invest in biotechnological solutions. Additionally, there are clinical or subclinical Retinol deficiencies in nations like Asia, Africa, and Latin America. The carotenes, especially β-carotene, which are found in many fruits and vegetables, are the predominant sources of vitamin A. Poor individuals rarely have access to inexpensive, high-carotene fruits and vegetables, and those who can afford them do not always use them as a food source. Therefore, the gene phytoene synthase has been added to canola and mustard stores, and it has been shown that when this gene is expressed, there is a significant buildup of β-carotene in the canola mustard seeds, hence fulfilling the nutritional requirement of these people (Kishore and Shewmaker 1999). Researchers have introduced cystathionine γ-synthase (CGS) to legumes, which is a critical enzyme in methionine synthesis. Transgenic alfalfa plants expressing *Arabidopsis* CGS demonstrated a 32-fold increase in soluble methionine level compared to wild-type plants (Avraham et al. 2005). Golden rice, modified to produce provitamin A (β-carotene), exemplifies biotechnology's promise to address this deficit. Transgenic lines now produce up to 35 μg/g of carotenoids, a substantial increase from previous levels (Paine et al. 2005). Remarkable progress has been in improving vital amino acids, altering starch composition, raising healthy fatty acids and vitamins, and increasing mineral levels in key staple crops. With continuous progress of biotechnological techniques, together with conventional breeding, shows enormous potential for solving global nutritional inadequacies by producing more nutritious plant-based foods and feeds (Dutta and Chatterjee 2024). The most extensively cultivated grain crop, maize (*Zea mays*), is used for a variety of purposes, including animal feed, biofuel, and various human nutritional products like corn syrup and starch. When ingested by animals and humans, phytate, also termed inositol 1,2,3,4,5,6-hexakisphosphate, acts as an anti-nutrient, reducing protein and mineral absorption by creating insoluble complexes with them (Feil et al. 2005). Zinc finger nucleases (ZFNs) were used to inhibit the IPK1 gene, which encodes inositol-1,3,4,5,6-pentakisphosphate 2-kinase, reducing phytate concentration in corn (Ku and Ha 2020).

Sulfur-containing amino acids methionine and cysteine are present in lesser quantities in legumes, root, tuber, and most vegetable proteins. Several molecular techniques have been tried and developed to remedy such deficiencies either by expanding the pool of a certain free essential amino acids (EAA) by metabolic engineering or by increasing the content of protein-bound EAAs in crops (Sun and Liu 2004). To increase the content of methionine and lysine in rice, the modified Met-rich soybean glycinin gene under the control of the GluB-1 promoter was transformed and expressed in rice, amounting to 5% of total protein. But the modified soybean Met-rich glycinin could not only assemble into trimeric and hexameric structures in rice as those observed in developing soybean seeds, but also rice glutelin could assemble with the modified soybean to form soluble hetero-trimers and hexamers and insoluble complexes in the rice seeds (Katsube et al. 1999; Samuel 2008). Although several alterations in genes of many crops have been done, still a lot can be done to fulfill the dietary requirements at affordable expenditure in future.

6.2.3 Postharvest Physiology

In order to prolong shelf life and maintain quality, postharvest physiology studies the physiological alterations that take place in plant tissues following harvest. It is critical for knowing how fruits, vegetables, and other foods decay, and how to slow these processes down through various storage and handling approaches. Various physiological processes (such as respiration, transpiration, ethylene production, and senescence) and physical factors (such as temperature, humidity, light, atmospheric composition, and handling) are key primary drivers of change in postharvest physiology of fruits, vegetables, and other plant products, leading to their deterioration (Brummell, Bowen, and Gapper 2022). Such deterioration involves changes in color, texture, and flavor. Understanding this effect on these products can help in mitigation of these changes. For example, storing these products at low temperature, modified atmosphere (low oxygen, high carbon dioxide, high humidity, and lesser light exposure) resulting in less transpiration and ethylene production, gentle handling, antimicrobial agents (reduce the risk of microbial spoilage), good packaging materials, and genetically modified crops can slow down the postharvesting alterations in plant-based products (Sharma and Kumari 2021). Postharvest decay is a severe concern for vegetables worldwide, and the extent of postharvest loss due to rot is widely documented. Pathogens are thought to decompose approximately 20%–25% of harvested vegetables in wealthy nations during postharvest processing (Sharma, Singh, and Singh 2009; Singh and Sharma 2018; Droby 2006; El Ghaouth, Wilson, and Wisniewski 2004). Because of inadequate storage, processing, and transit facilities, postharvest decay rates in poorer nations are quite high (over 35%). Harmful fungicides (Benomyl and Iprodione) that are used to control the postharvest diseases of vegetables, have adverse impacts on the health of the consumers (El Ghaouth, Wilson, and Wisniewski 2004). Chilling damage is a series of alterations brought on by storing fruits at low temperatures that affects the quality of vegetable products and as a result, lowers their economic value. Membrane damage and reactive oxygen species are the primary cold injury causes that harm delicate horticultural products. Chilling injury tolerance is due to certain plant species' ability to store heat shock proteins (HSPs). Hence, biotechnological techniques play an important role in the expression of HSPs leading to chilling tolerance during postharvest cold storage (Sevillano et al. 2009). HSP70, chaperonins (HSP60), HSP90, and HSP100 are the identified families of HSPs with molecular weights ranging from 15 to 115 kD present in cytoplasm, nucleus, mitochondria, chloroplast, and endoplasmic reticulum (Wang et al. 2004; Timperio, Egidi, and Zolla 2008). Yeast SAMDC gene under the control of E8 promotor showed improvement in lycopene content of tomato along with better quality of vine life and fruit juice without compromising fruit softness and color (Bapat et al. 2010). Transgenic fruits had a shelf life of up to 60–120 days (e.g., tomato) at room temperature with no change in their firmness or color (Rose et al. 2003). One of the technologies used to postpone postharvest changes and increase the fruits' and vegetables' shelf life is RNA interference. Gene editing, clustered regularly interspaced short palindromic repeat/CRISPR-associated protein 9 (CRISPR/Cas9), and targeting induced local lesions in genomes are examples of transgenic technologies that have been successfully used in a variety of species to increase productivity, develop tolerance to biotic and abiotic stresses, increase shelf life, and boost nutritional quality. Globally, transgenic crops—also referred to as genetically modified crops, or GM crops—are cultivated, including cassava, soybean, brinjal, and tomato. These strategies are very promising in establishing food security, improving crop yield, reducing postharvest losses, and producing secondary metabolites, hormones, and plantibodies (Mihrete, Workie, and Bogale 2024). Modified atmosphere packaging (MAP) is one of the most widely used new technologies that involves changing the composition of gases surrounding agricultural items (Jones and Patel 2023). MAP tries to reduce the respiration rate and prolong the product's senescence by raising carbon monoxide levels while decreasing oxygen levels (Jones and Patel 2021). Breeding for better postharvest quality in Zucchini and *Brassica* using CRISPR/Cas9-induced ET mutations. These findings confirmed the hypothesis that CRISPR/Cas9-induced alterations in the tomato ripening TF did not completely inhibit ripening (Ninama et al. 2024).

6.2.4 Genome Editing Technologies

Continuous rise in the human population and reduction of agricultural land pose a great challenge of ensuring food security for humans. Therefore, new methods have been developed to improve the crops precisely, in an ecologically safer and faster manner (Pixley et al. 2022). A wide range of tools are available for precise editing of plant genomes (Kumar et al. 2023). Genome editing offers benefits such as nutritional enhancement; better food safety; reduced food waste; resistance to diseases, weeds, and pests; increased seed affordability due to cheaper seed manufacturing;

enhanced climatic resilience, particularly drought tolerance; and increase in biodiversity (Sedeek, Mahas, and Mahfouz 2019; Eshed and Lippman 2019). Targeted genome editing technologies, particularly CRISPR/Cas systems, have transformed basic research and crop breeding by allowing precise targeted modification of an organism's DNA (Mishra, Joshi, and Zhao 2022). Site-directed nucleases (type 1 and type 2) are used to make double-strand break (DSB) at random sites to delete or add nucleotides, leading to improvements such as climate resilience in rice; disease resistance in yam, potato, banana, wheat, and maize; nutrition improvement in cassava; digestibility in bean, high provitamin A in maize; reduced glycemic index and apomixis in rice and potato; heat tolerance in potato; and reduced aflatoxin in groundnuts (Pixley et al. 2022). Site-specific recombinase (SSR) is a frequent genetic engineering approach that causes a permanent alteration to the targeted genome. Several SSR systems, including Cre/loxP and Flp/FRT, have been identified to accomplish DNA rearrangements (Allen and Weeks 2009; Araki et al. 1995; Allen and Weeks 2005; Yueju Wang et al. 2011). SSRs are useful tools for modifying genomes and activating or deactivating gene expression in many organisms (Abdallah, Prakash, and McHughen 2015). SSR or site-specific nuclease (SSN) mechanisms are used in a wide range of tools to carry out genome editing. Gene swapping could be done by using SSR via tag and exchange method. The first step involves homologous recombination to integrate a SSR cassette with a marker gene in the target site, followed by cassette exchange. The tag stage necessitates a long homologous sequence, whereas SSRs needs a small homology of roughly 50 kb (Abdallah, Prakash, and McHughen 2015). By causing knockout or knock-in mutations in the eukaryotic genomes, recombinase has been widely used to alter DNA in bacteria, yeast, plants, and humans. Recombinases show numerous merits like they are not dependent on intercellular repair mechanisms. Furthermore, they have the capability to perform their function without forming DSBs. Additionally, they insert donor DNA into genome. But their chimeric enzymes have significant off-target effects and toxicity when expressed at large levels (Anastassiadis et al. 2009). Site-specific endonucleases, which can cleave DNA in a specific sequence within the genome, are necessary for nucleases to function. A DNA-binding domain or RNA sequence found in SSN attaches itself to the target sequence. Endonucleases, which can cleave DNA in a specific sequence inside the genome, are essential for SSNs. SSN has a DNA-binding domain or RNA sequence that binds to the target sequence and cleave it, followed by DNA repair mechanism of the cell leading to modification of gene at targeted site (Gaj, Gersbach, and Barbas 2013; Carroll 2014). Engineered meganuclease (MegaN), ZFNs, transcription activator-like effector nucleases (TALENs), and CRISPR/Cas9 nuclease system are four major families of nucleases used for genome editing, which enable us to change the genome in an easy, efficient, and cost-effective manner (Čermák et al. 2015; Voytas 2013). By moving to the nucleus and causing DNA DSBs at specific endogenous genomic sites, programmable SSNs enable precise genome editing. Breaks using various SSN platforms trigger DNA repair pathways in the target cell, including deletions and transgene insertions (Abdallah, Prakash, and McHughen 2015). By deleting, inserting, or changing certain DNA sequences, the novel CRISPR-Cas9 technology allows geneticists and medical researchers to modify portions of the genome. It is currently the simplest, most versatile and precise method of genetic manipulation. Several species have shown increased resistance to fungal diseases by focusing on homologs of Mildew resistance locus (MLO) and other loci. By simultaneously targeting the three homologs of the Mildew resistance locus (MLO), TaMLO-A, TaMLO-B, and TaMLO-D, in wheat, CRISPR/Cas9 and TALEN were able to successfully create resistance to powdery mildew (Yanpeng Wang et al. 2014). Important staple crops can suffer greatly from plant viruses, and some plant virus epidemics can have a severe negative economic impact. Interference against these DNA viruses using CRISPR/Cas9 is quite beneficial to make plants immune against these plants. To create better foods, genome editing improves the nutritional qualities of crops. Numerous research studies have shown possible uses for genome editing in altering plant components. For example, acrylamide is a possible carcinogen that is created when high heat reacts reducing sugars (like glucose and fructose) with free amino acids (like asparagine) in starchy foods like potatoes. Clasen et al. (2016) created acrylamide-free potatoes by using TALEN to knock out VInv, the gene for vacuolar invertase, which catalyzes the breakdown of sucrose into glucose and fructose. Many crops contain phytate, which is typically considered an anti-nutrient because it can combine with proteins and minerals to form complexes that decrease their availability for digestion (Zhou and Erdman 1995). By deleting ZmIPK, a gene involved in phytate production, TALENs and CRISPR/Cas9 have both been used to lower the amount of phytate in maize (Liang et al. 2014). By creating functional knockout mutants of the α-gliadin gene and multiple genes involved in carotenoid biosynthesis, CRISPR/Cas9 has been utilized to produce wheat with hypoimmunogenic gluten and tomatoes with increased lycopene content (Li et al. 2018; Sánchez-León et al. 2018).

6.2.5 Global Food Security and Economic Opportunities

Technology has acted as gamechanger in the food system. In recent decades, biotechnology has transformed our food system by providing creative answers to urgent issues such as food security, nutritional quality, and environmental sustainability. By leveraging the power of genetic engineering, tissue culture, molecular biology, and other cutting-edge techniques, it has become possible to develop crops with improved disease and pest resistance, nutritional profile, and yields, transforming the way we produce, distribute, and consume food on a global scale. Understanding the role and possibilities of biotechnology in changing the future of our food supply is more important than ever as we navigate the complexity of feeding a growing population while limiting our environmental impact. Biotechnology has enabled us to produce higher yields of crops for supplying to the growing population, especially in the developing countries. By raising agricultural yields and lowering crop loss, biotechnology can increase global food production and decrease food insecurity. Farmers in emerging countries must, like any other technological advancement, find the new developments beneficial. Consumers in developing nations will profit from biotech crops that are less expensive or more nutritious than traditional ones (Lorraine 2001). Biofortification is used to combat hidden hunger. Instead of adding vitamins and minerals during processing, biofortification involves crossbreeding commonly produced crop varieties with carefully selected seeds from gene banks with naturally increased amounts of iron, zinc, or vitamin A. Biofortified staple crops are a practical and cost-effective solution, giving familiar and healthy options for smallholder farming households that rely on what they raise themselves (Areche et al. 2023). Improved yields and quality of crops are not only fulfilling the needs of growing human population, but also providing good income to the farmers. Techniques used in biotechnology such as crossbreeding, and gene modification by CRISPR/Cas9 have been found beneficial for the mankind (Qaim and Virchow, n.d.; Najafi and Lee 2014; Tait and Barker 2011; Timmer 2003).

6.3 OPPORTUNITIES IN ANIMAL BIOTECHNOLOGY INNOVATION ACTION

Animal biotechnology uses molecular biology techniques to genetically modify animals for a variety of goals, including increased agricultural production, pharmaceutical development, and improved animal health and welfare. This field involves a wide range of technologies, including traditional breeding methods and modern genetic engineering techniques such as CRISPR. Animal biotechnology is a branch that modifies animals genetically, that is, by editing their genomes using molecular biology techniques to make them more suitable for use in industrial, medicinal, or agricultural settings. Animal biotechnology plays a fundamental role in improving the animal health and immunity against diseases, enhancing productivity, pharmaceutical industry, and creating transgenic animals by different techniques, leading to the welfare of the society. Biotechnology has shown its applications in diverse areas as discussed ahead.

6.3.1 Improved Animal Health and Welfare

Biotechnology has significantly influenced the animal health and welfare through a variety of uses, including increasing feeding, generating vaccinations and diagnostic tools, and fostering disease resistance. These innovations contribute to more sustainable and ethical livestock practices, while also helping human health with safer and more nutritious food products (Gohar et al. 2022). Biotechnology can enhance animal performance by enhancing nutrition, output potential, and health status. Nutrients, like as amino acids, can be created and/or safeguarded to improve food formulation and satisfy specific productive demands. Enzymes can improve nutrient availability from feedstuffs, minimize feed costs, and reduce waste output to the environment (Bonneau and Laarveld 1999). Biotechnological techniques are used in diagnostic and pharmaceutical industry. Therapeutic proteins can be produced by genetically modified organisms (GMOs). The invention of recombinant insulin in the 1980s, made using genetically altered *Escherichia coli* bacteria, was a key milestone in biotech commercialization, delivering a safer and more reliable source of insulin for millions of diabetics worldwide (Bifulco et al. 2025). CRISPR/Cas9 gene editing technology has transformed genetic engineering by introducing the potential to alter an organism's DNA for precision medicine. This groundbreaking technique enables scientists to precisely modify DNA, opening up new avenues for treating genetic illnesses and developing biotechnology (S. Kumar et al. 2024).

Biotechnology is not just restricted to research for human welfare, but also for other animals to improve the quality of the products that we got from them. Precise genetic alterations are possible with CRISPR/Cas9, stem cell-based regenerative therapies, and advanced molecular diagnostics which are already being employed to increase disease resistance, boost animal productivity, and aid in wildlife conservation. To protect animal welfare and environmental safety, these developments do, however, present ethical and legal issues that need to be carefully handled. It is crucial that biotechnology be applied responsibly, under strict guidelines and with ethical concerns. By facilitating accurate diagnosis and customized therapies, cutting-edge technologies like artificial intelligence and personalized medicine hold the potential to further transform veterinary care in the future (Kumar et al. 2024). Biotechnologies like artificial insemination, embryo transfer, sexing, cloning, and other related processes might impact the effectiveness of reproduction and, consequently, selection programs (Said et al. 2020). Livestock sector is more prominent than agricultural output in the developed countries. It is growing at faster rate than agriculture sector and will grow further at higher rates than agricultural sector. The output from meat animals is three times faster than that of growth in animals and number of dairy animals (Cunningham 1999). With increase in population and limited resources, the need of livestock will also increase. Therefore, biotechnology will be crucial to achieving this. Artificial insemination is widely used in commercial dairy farming. Semen from higher genetic merit is used to produce the progeny with higher milk output (Cunningham 1999). Furthermore, recombinant somatotropin (ST) is administered, meat animals grow more quickly and produce slimmer carcasses, while dairy cows produce more milk. There are few effects on the quality of milk or meat. Additionally, immunomodulation can be employed to increase endogenous anabolic hormone activity. The effects of transferring the ST gene are identical to those of using exogenous ST. However, ad-hoc promoters are used to regulate the gene's expression, but the transgenic animals' health is significantly compromised (Bonneau and Laarveld 1999). Similar is the case of other meat-producing animals. They are injected with growth hormones to accelerate the growth of these animals in poultry, mutton, and pork industry to get desirable growth of the livestock (Pashen 1987). The dairy industry needs a lot of rennet to make cheese in vast quantities. Rennet is traditionally obtained from calves. However, due to the global imbalance between calf slaughter and cheese production, as well as the global scarcity of calf rennet, it is necessary to look into other sources of rennet. The demand for renn*et* alternative has been made worse by religious sentiment in some nations, such as India. Various enzymes, especially galactosidase, are produced commercially and economically from yeasts (*Kluyveromyces lactis* and *Kluyveromyces marxianus*) by using biotechnological techniques to solve the problem of lactose intolerance in humans (Gupta and Savalia 2012). Products made from livestock are extremely vulnerable to microbial contamination and degradation. Therefore, these are preserved by using lactic acid bacteria naturally, as they have the antagonistic properties to a wide range of bacteria like *Salmonella, Staphylococcus, Clostridium*, *Yersinia enterocolitica*, and *Pseudomonas* instead of using harmful chemical preservatives (Sarkar and Mishra 2001; Gálvez et al. 2014).

6.3.2 Benefits for Human Health and the Environment

Biotechnology has played an imperative role in vaccine development, as evidenced by the prompt response to the COVID-19 pandemic. Pfizer and Moderna, two major biotech companies, earned attention for their mRNA-based vaccines. These vaccines constituted a significant leap in vaccination technology, allowing for speedier development and customization (Gohar et al. 2022). Combining DNA molecules from two different species and transferring the resulting DNA into a host organism is known as recombinant DNA technology. New genetic combinations for industry, agriculture, and medicine will be produced by that host organism. Applications for recombinant DNA technology are numerous and include agricultural biotechnology, which includes modified fruits and vegetables, biopharmaceuticals and diagnostics, and energy applications, such as biofuel. Genetically modified items can outperform conventional medicines and produce. Recombinant agriculture can be more pest- or weather-resistant; recombinant pharmaceuticals, such as insulin, can operate more effectively with the body, and so on (Singh 2018). Another extraordinary biotech success story involves monoclonal antibodies (mAbs). These designed proteins have emerged as a cornerstone of modern medicine, providing precise treatment for a variety of ailments. Herceptin (trastuzumab), created by Genentech, shows the promise of mAb. By selectively targeting cancer cells, Herceptin, which received FDA approval in September 1998, transformed the treatment of HER2-positive breast cancer, while sparing healthy ones. Its success not only helped patients live longer lives, but also paved the way for a new era of personalized treatment (Gohar et al. 2022). Biotechnology has also made genome sequencing accessible, making it more accessible and cost-effective than before. The Human Genome Project was a remarkable accomplishment in reading the entire human genetic code. This research paved the way for personalized therapy and identifying genetic markers linked to diverse diseases (Afzal et al. 2016). Xenotransplantation is the process of grafting or transferring organs or tissues between members of other species. The increasing number of patients with end-stage heart disease and the lack of available donors have sparked new interest in the prospect of cardiac xenotransplantation. Benefits include expanding the pool of potential donors, preventing the detrimental effects of brain death on donor organ function, and having an infinite supply of donor organs available for elective use. The pig is the only animal that can replace human tissues because of ethical and ecological restrictions, even though baboons and monkeys are the most phylogenetically related to humans. The pig can be reared in controlled conditions and, more significantly, genetically changed since its organ sizes and physiological characteristics are similar to those of people. The primary disadvantages of xenotransplantation are the potential for rejection, including hyperacute, acute, and chronic vascular, and the spread of zoonoses. The creation of genetically altered pigs has made it possible to express human antigens, which can lower the likelihood of rejection and give hope for future viability (Weber, Kobashigawa, and Givertz 2017). Besides the aforementioned applications of biotechnology, there are other numerous useful techniques which are used for improving animal as well as human welfare. There are a number of products which are invented at regular basis and introduced into the market, which revolutionized the industry drastically,

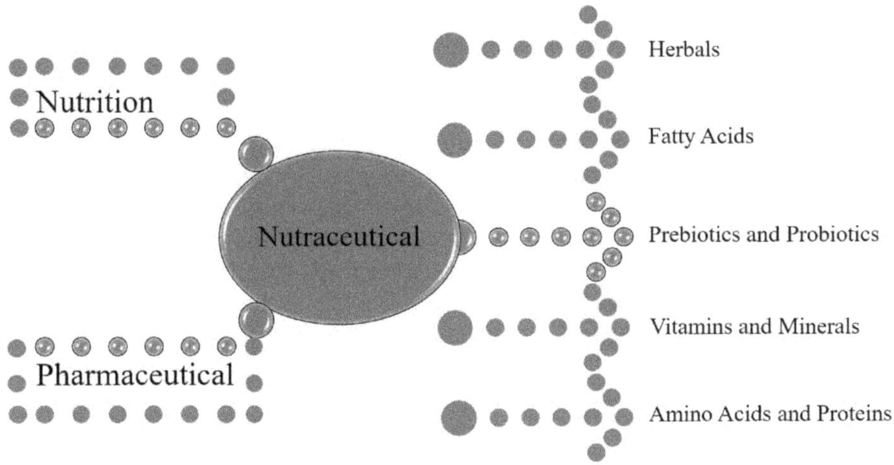

FIGURE 6.2 Uses of biotechnology in synthesis of bioproducts.

leading to the growth of the market and prove to be profitable invention. Transgenic animals and cloning are also a big breakthrough of biotechnology. Both these techniques are very useful either to produce new traits in an individual by transferring the genes from other species or to produce genetically identical copies of animals for research, or to preserve endangered species by using genetic material of the individual. Hence, biotechnology plays an imperative role in human population, right from fulfilling the dietary requirements, to keep them disease-free with the help of advancements in pharmaceutical products by using different cytological and molecular techniques (Figure 6.2).

6.4 CHALLENGES

6.4.1 Regulatory Hurdles

One of the most difficult parts of creating and launching biotech goods is navigating the regulatory environment. Product safety, efficacy, and quality are guaranteed by regulatory compliance; nevertheless, comprehending and fulfilling these standards can be difficult and time-consuming. This article provides a comprehensive guide on overcoming regulatory barriers in the biotechnology industry, along with strategies and insights to assist biotechnology startups and established enterprises succeed. Key challenges in biotech products are complex regulations for the products involving genetic material or medicine due to their novelty and lengthy regulatory processes from authorities like FDA or regional regulatory authorities for approval.

6.4.2 Funding and Financial Pressures

Despite its revolutionary potential, biotechnology is severely limited by the high expenses of research, development, and regulation. Funding choices, project prioritization, and the general rate of innovation can all be impacted by these difficulties. It can take years and cost millions to develop new biotech products, such as medications or diagnostics, because they require a great deal of scientific testing, validation, and experimentation. It can take up to 10 years to bring a new biotech product to market, which means there will not be any income for a while. Growing biotech businesses frequently necessitates large infrastructure, equipment, and facility investments. Additionally, clinical studies are quite expensive, and unanticipated setbacks or delays may result in cost overruns. The expense of developing biotech products is increased by the strict regulatory criteria they must meet. Due to their lack of experience, early-stage businesses may find it difficult to obtain finance, and mature businesses are up against more competition for capital. Biotech development carries a significant chance of failure, which might put off investors, especially those looking for rapid returns. Ventures with clearer routes to market and significant revenue potential may be prioritized due to financial constraints, possibly at the price of riskier but more creative ventures.

6.4.3 Ethical Dilemmas

Bioethics has both historical and contemporary uses in biotechnology. The most popular biotechnology field is the one that deals with medicine and other related fields. The field is also known as red biotechnology. Biotechnology performs a number of experiments and investigations, some of which are unethical, to achieve satisfying findings. Particular problems that are immoral are known as ethical problems in biotechnology. These kinds of problems can have an impact on the core moral principles that cause a great deal of friction with society's regular operations. There are numerous ethical concerns which act as constraints in the research and development of new products such as environmental, legal, cultural, and socioeconomic issues. Besides these, many other ethics are part of this industry.

6.5 CONCLUSION

A strategy framework for using biotechnology's potential to promote agriculture, food security, sustainability, and economic growth is provided by the Plant and Animal Biotechnology Innovation Action Plan. Improving crop yields, strengthening resilience to pests and diseases, increasing animal output, lessening the impact on the environment, and tackling global issues like food shortages and climate change are the main opportunities. Precision breeding, quicker genetic advancements, and the creation of new features that were previously impossible with conventional techniques can all be made possible by biotechnology. But there are also big obstacles in the way of these chances. Public acceptance, possible threats to biodiversity, ethical and regulatory challenges, and unequal access to biotechnological developments across areas are some of the main concerns. Overcoming these obstacles requires ensuring open governance, building public confidence, and encouraging global cooperation. Although plant and animal biotechnology have the potential to be revolutionary, its appropriate use and long-term success depend on a fair and inclusive strategy based on innovation, safety, and sustainability.

REFERENCES

Abdallah, Naglaa A., Channapatna S. Prakash, and Alan G. McHughen. 2015. "Genome Editing for Crop Improvement: Challenges and Opportunities." *GM Crops & Food* 6(4): 183–205. doi:10.1080/21645698.2015.1129937.

Abraham, Binju, Hailu Araya, Tareke Berhe, Sue Edwards, Biksham Gujja, Ram Bahadur Khadka, Yang S Koma, et al. 2014. "The System of Crop Intensification: Reports from the Field on Improving Agricultural Production, Food Security, and Resilience to Climate Change for Multiple Crops." *Agriculture and Food Security* 3(4): 1. doi:10.1186/2048-7010-3-4.

Afzal, Hina, Khadija Zahid, Qurban Ali, Kubra Sarwar, Sana Shakoor, Ujala Nasir, and Idrees Ahmad Nasir. 2016. "Role of Biotechnology in Improving Human Health." *Journal of Molecular Biomarkers & Diagnosis* 7(6). doi:10.4172/2155-9929.1000309.

AL-Eitan, Laith, and Malek Alnemri. 2022. "Biosafety and Biosecurity in the Era of Biotechnology: The Middle East Region." *Journal of Biosafety and Biosecurity* 4(2):130–145. doi:10.1016/j.jobb.2022.11.002.

Allen, Bryan G., and Daniel L. Weeks. 2005. "Transgenic Xenopus Laevis Embryos Can Be Generated Using ΦC31 Integrase." *Nature Methods* 2(12): 975–979. doi:10.1038/nmeth814.

Allen, Bryan G, and Daniel L Weeks. 2009. "Bacteriophage ΦC31 Integrase Mediated Transgenesis in Xenopus Laevis for Protein Expression at Endogenous Levels." In *Microinjection: Methods and Applications*, edited by David J Carroll, pp. 113–122. Totowa, NJ: Humana Press. doi:10.1007/978-1-59745-202-1_9.

Anastassiadis, Konstantinos, Jun Fu, Christoph Patsch, Shengbiao Hu, Stefanie Weidlich, Kristin Duerschke, Frank Buchholz, Frank Edenhofer, and A. Francis Stewart. 2009. "Dre Recombinase, like Cre, Is a Highly Efficient Site-Specific Recombinase in *E. Coli*, Mammalian Cells and Mice." *DMM Disease Models and Mechanisms* 2(9–10): 508–515. doi:10.1242/dmm.003087.

Araki, Kimi, Masatake Araki, Jun-Ichi Miyazaki, and Pierre Vassalli. 1995. "Site-Specific Recombination of a Transgene in Fertilized Eggs by Transient Expression of Cre Recombinase (Microinjection/Chicken J3-Actin Promoter/(3-Galactosidase)." *Proceedings of the National Academy of Sciences* 92(1): 160–164.

Areche, F. O., A. H. Gondal, L. A. Sumarriva-Bustinza, N. O. Zela-Payi, J. M. Sumarriva-Bustinza, R. B. Oscanoa-León, A. F. Calcina-Sotelo, et al. 2023. "Role of Biotechnology in Food Security: A Review." *Sabrao Journal of Breeding and Genetics* 55(5): 5. doi:10.54910/sabrao2023.55.5.5.

Avraham, Tal, Hanna Badani, Shmuel Galili, and Rachel Amir. 2005. "Enhanced Levels of Methionine and Cysteine in Transgenic Alfalfa (*Medicago Sativa* L.) Plants over-Expressing the Arabidopsis Cystathionine γ-Synthase Gene." *Plant Biotechnology Journal* 3(1): 71–79. doi:10.1111/j.1467-7652.2004.00102.x.

Banga, Surinder Singh., and Manjit Sing. Kang. 2014. "Developing Climate-Resilient Crops." *Journal of Crop Improvement* 28(1): 57–87. doi:10.1080/15427528.2014.865410.

Bapat, Vishwas A., Prabodh K. Trivedi, Antara Ghosh, Vidhu A. Sane, Thumballi R. Ganapathi, and Pravendra Nath. 2010. "Ripening of Fleshy Fruit: Molecular Insight and the Role of Ethylene." *Biotechnology Advances* 28(1): 94–107. doi:10.1016/j.biotechadv.2009.10.002.

Bhatia, Saurabh. 2015. "History and Scope of Plant Biotechnology." In *Modern Applications of Plant Biotechnology in Pharmaceutical Sciences*, edited by Saurabh Bhatia, Kiran Sharma, Randhir Dahiya, and Bera Bera, 1st ed., pp. 1–30. Amsterdam: Elsevier. doi:10.1016/B978-0-12-802221-4.00001-7.

Bhojwani, S. S., and M. K. Razdan. 1996. *Plant Tissue Culture: Theory and Practice*, 1st ed., vol. 5. Amsterdam: Elsevier.

Bifulco, Maurizio, Erika Di Zazzo, Alessandra Affinito, and Cristina Pagano. 2025. "The Relevance of the History of Biotechnology for Healthcare: Teaching Students How Biotechnology and Medicine Have Been Closely Entwined during the Past Century Highlights How Both Fields Have Inspired and Driven Each Other." *EMBO Reports* 26: 303–306. doi:10.1038/s44319-024-00355-8.

Bonneau, M, and B Laarveld. 1999. "Biotechnology in Animal Nutrition, Physiology and Health." *Livestock Production Science* 59(2–3): 223–241.

Brummell, David A, Joanna K Bowen, and Nigel E Gapper. 2022. "Biotechnological Approaches for Controlling Postharvest Fruit Softening." *Current Opinion in Biotechnology* 78(December): 102786. doi:10.1016/j.copbio.2022.102786.

Carroll, Dana. 2014. "Genome Engineering with Targetable Nucleases." *Annual Review of Biochemistry* 83: 409–439. doi:10.1146/annurev-biochem-060713-035418.

Čermák, Tomáš, Nicholas J. Baltes, Radim Čegan, Yong Zhang, and Daniel F. Voytas. 2015. "High-Frequency, Precise Modification of the Tomato Genome." *Genome Biology* 16(1). doi:10.1186/s13059-015-0796-9.

Chauhan, Aman Pratap Singh Chauhan, Dheerendra Singh, Om Prakash Sharma Sharma, Nishita Kushwah, and Alpana Kumhare. 2023. "Agronomic Practices for Enhancing Resilience in Crop Plants." *Plant Science Archives* 8(3): 1–3. doi:10.51470/PSA.2023.8.3.01.

Clasen, Benjamin M., Thomas J. Stoddard, Song Luo, Zachary L. Demorest, Jin Li, Frederic Cedrone, Redeat Tibebu, et al. 2016. "Improving Cold Storage and Processing Traits in Potato through Targeted Gene Knockout." *Plant Biotechnology Journal* 14(1): 169–176. doi:10.1111/pbi.12370.

Cunningham, E. P. 1999. "The Application of Biotechnologies to Enhance Animal 1 Production in Different Farming Systems." *Livestock Production Science* 58(1): 1–24.

Droby, S. 2006. "Improving Quality and Safety of Fresh Fruits and Vegetables After Harvest by the Use of Biocontrol Agents and Natural Materials." In *International Symposium on Natural Preservatives in Food Systems*, edited by D. Havkin Frenkel, pp. 45–51. Leuven, Belgium: Acta Horticulture.

Dutta, Angira, and Avishek Chatterjee. 2024. "Biotechnological Approach to Improve Nutritional Quality of Plants." *International Journal of Agriculture and Nutrition* 6(2): 71–74. doi:10.33545/26646064.2024.v6.i2b.182.

Eshed, Yuval, and Zachary B. Lippman. 2019. "Revolutions in Agriculture Chart a Course for Targeted Breeding of Old and New Crops." *Science* 366(6465): 880–884. doi:10.1126/science.aax0025.

Food and Agriculture Organization, World Food Programme, and International Fund for Agricultural Development. 2012. *The State of Food Insecurity in the World*. Rome: Food and Agriculture Organization.

Farre, Gemma, Sonia Gomez-Galera, Shaista Naqvi, Chao Bai, Georgina Sanahuja, Dawei Yuan, Uxue Zorrilla, et al. 2012. "Biotechnology and Nutritional Improvement of Crops." In *Encyclopedia of Sustainability Science and Technology*, edited by Robert A Meyers, pp. 1676–1723. New York, NY: Springer. doi:10.1007/978-1-4419-0851-3_160.

Feil, B, S B Moser, S Jampatong, and P Stamp. 2005. "Mineral Composition of the Grains of Tropical Maize Varieties as Affected by Pre-Anthesis Drought and Rate of Nitrogen Fertilization." *Crop Ecology, Management & Quality* 45: 516–523.

Gaj, Thomas, Charles A. Gersbach, and Carlos F. Barbas. 2013. "ZFN, TALEN, and CRISPR/Cas-Based Methods for Genome Engineering." *Trends in Biotechnology*. doi:10.1016/j.tibtech.2013.04.004.

Gálvez, Antonio, Rosario Lucas López, Rubén Pérez Pulido, and María José Grande Burgos. 2014. "Biopreservation of Milk and Dairy Products." In *Food Biopreservation* (*SpringerBriefs in Food, Health, and Nutrition*), pp. 49–69. New York: Springer. doi:10.1007/978-1-4939-2029-7_5.

Ghaouth, Ahmed El, Charles Wilson, and Michael Wisniewski. 2004. "Biologically-Based Alternatives to Synthetic Fungicides for the Control of Postharvest Diseases of Fruit and Vegetables." In *Diseases of Fruits and Vegetables: Volume II: Diagnosis and Management*, edited by S A M H Naqvi, pp. 511–535. Dordrecht: Springer. doi:10.1007/1-4020-2607-2_14.

Gohar, Umar Farooq, Javaria Sarwar, Hiba Akram, and Hamid Mukhtar. 2022. "Recent Advances in Biotechnology in Animal Health." In *Veterinary Pathobiology and Public Health*, 1st ed., pp. 511–533. Faisalabad, Pakistan: University of Agriculture Press. doi:10.47278/book.vpph/2021.044.

Gupta, Swati, and C. V. Savalia. 2012. "Application of Biotechnology to Improve Livestock Products." *Veterinary World* 5(9): 571–574. doi:10.5455/vetworld.2012.634-638.

Jha, Timir Baran, and Biswajit Ghosh. 2005. *Plant Tissue Culture: Basic and Applied*, 1st ed., vol. 1. Hyderabad: University Press.

Jones, A, and B. Patel. 2021. "Emerging Technologies for Enhancing the Shelf Life of Horticultural Crops." *Journal of Postharvest Technology* 15(2): 123–135.

Jones, A., and B. Patel. 2023. "Genetic Manipulation of Cytokinin Levels for Delaying Leaf Senescence in Leafy Vegetables." *Journal of Agricultural Biotechnology* 12(3): 45–57.

Katsube, Tomoyuki, Nobuyuki Kurisaka, Masahiro Ogawa, Nobuyuki Maruyama, Reiko Ohtsuka, Shigeru Utsumi, and Fumio Takaiwa. 1999. "Accumulation of Soybean Glycinin and Its Assembly with the Glutelins in Rice 1." *Plant Physiology* 120(4): 1063–1074.

Kishore, Ganesh M, and Christine Shewmaker. 1999. "Colloquium Paper This Paper Was Presented at the National Academy of Sciences Colloquium "Plants and Population: Is There Time?" *Proceedings of the National Academy of Sciences of the United States of America* 96: 5968–5972.

Ku, Hyung Keun, and Sun Hwa Ha. 2020. "Improving Nutritional and Functional Quality by Genome Editing of Crops: Status and Perspectives." *Frontiers in Plant Science* 11: 577313. doi:10.3389/fpls.2020.577313.

Kumar, Rahul, Ashutosh Sharma, Yogesh Kumar, and Sharma Indu. 2023. *Genome Editing for Crop Improvement*, 1st ed., vol. 1. New York, NY: Nova Science Publishers. doi:10.52305/RZCJ7386.

Kumar, Sudesh, Suvidhi, B. C. Bera, and R. K. Vaid. 2024. "The Future of Veterinary Medicine: Integrating Biotechnology for Improved Animal Health and Welfare." *Journal of Animal & Plant Husbandry* 4(1): 180021. https://academicstrive.com/submit-manuscript.php.

Li, Xindi, Yanning Wang, Sha Chen, Huiqin Tian, Daqi Fu, Benzhong Zhu, Yunbo Luo, and Hongliang Zhu. 2018. "Lycopene is Enriched in Tomato Fruit by CRISPR/Cas9-Mediated Multiplex Genome Editing." *Frontiers in Plant Science* 9(April): 559. doi:10.3389/fpls.2018.00559.

Liang, Zhen, Kang Zhang, Kunling Chen, and Caixia Gao. 2014. "Targeted Mutagenesis in Zea Mays Using TALENs and the CRISPR/Cas System." *Journal of Genetics and Genomics* 41(2): 63–68. doi:10.1016/j.jgg.2013.12.001.

Liebman, Matt, and Lisa A. Schulte. 2015. "Enhancing Agroecosystem Performance and Resilience through Increased Diversification of Landscapes and Cropping Systems." *Elementa: Science of the Anthropocene* 3: 000041. doi:10.12952/journal.elementa.000041.

Liu, Chang, Daniel Plaza-Bonilla, Jeffrey A. Coulter, Hadley. Randy Kutcher, Hugh J. Beckie, Li Wang, Jean Baptiste Floch, et al. 2022. "Diversifying Crop Rotations Enhances Agroecosystem Services and Resilience." In *Advances in Agronomy*, edited by Donald Lewis Sparks, vol. 173, pp. 299–335. Cambridge, MA: Academic Press. doi:10.1016/bs.agron.2022.02.007.

Lorraine Mitchell. 2001. "Biotechnology and Food Security." *Issues in Food Security* 11: 761–762.

Mihrete, Tesfahun Belay, Melkamu Alemayehu Workie, and Fenta Assefa Bogale. 2024. "Applications of Biotechnology for Enhancing the Shelf Life of Horticultural Crops." *Journal of Sustainable Agriculture and Environment* 3: e12121. doi:10.1002/sae2.12121.

Mishra, Rukmini, Raj Kumar Joshi, and Kaijun Zhao. 2022. "Genome Editing Is Revolutionizing Crop Improvement." In *Genome Editing Technologies for Crop Improvement*, edited by Kaijun Zhao, Rukmini Mishra, and Raj Kumar Joshi, pp. 3–41. Singapore: Springer Nature. doi:10.1007/978-981-19-0600-8_1.

Mishra, S. P. 2009. *Plant Tissue Culture*, 1st ed., vol. 1. New Delhi: Ane Books Pvt Ltd.

Najafi, Mohammad, and Byong Lee. 2014. "Biotechnology and Its Impact on Food Security and Safety." *Current Nutrition & Food Science* 10(2): 94–99. doi:10.2174/1573401310666140306225243.

Narayanaswamy, B. N. 1994. *Plant Cell and Tissue Culture*, 1st ed., vol. 1. New Delhi: Tata McGraw-Hill Education.

Ninama, Nilesh, Lakshmi Gangal, Arshad Khayum, Harshitha S B, Divya C, Sowmya H M, Abhishek Singh, and Anushi. 2024. "Post-Harvest Biotechnology or Genetic Engineering Solutions: Extending Shelf Life and Reducing Food Waste." *Journal of Advances in Biology & Biotechnology* 27(4): 1–26. doi:10.9734/jabb/2024/v27i4736.

Paine, Jacqueline A., Catherine A. Shipton, Sunandha Chaggar, Rhian M. Howells, Mike J. Kennedy, Gareth Vernon, Susan Y. Wright, et al. 2005. "Improving the Nutritional Value of Golden Rice through Increased Pro-Vitamin A Content." *Nature Biotechnology* 23(4): 482–487. doi:10.1038/nbt1082.

Pashen, Robert. 1987. "Biotechnology: The Key to Improved Animal Production?" *Veterinary Clinics of North America: Food Animal Practice* 3(3): 647–656. doi:10.1016/S0749-0720(15)31136-1.

Pixley, Kevin V., Jose B. Falck-Zepeda, Robert L. Paarlberg, Peter W.B. Phillips, Inez H. Slamet-Loedin, Kanwarpal S. Dhugga, Hugo Campos, and Neal Gutterson. 2022. "Genome-Edited Crops for Improved Food Security of Smallholder Farmers." *Nature Genetics* 54(4): 364–367. doi:10.1038/s41588-022-01046-7.

Qaim, Matin, and Detlef Virchow. 2000. "The Role of Biotechnology for Global Food Security." *German Journal of Agricultural Economics* 49(9–10): 348–356.

Roberts, Daniel P., and Autar K. Mattoo. 2018. "Sustainable Agriculture—Enhancing Environmental Benefits, Food Nutritional Quality and Building Crop Resilience to Abiotic and Biotic Stresses." *Agriculture* 8: 8. doi:10.3390/agriculture8010008.

Rose, J. K. C., C Catalá, C. Z. H Gonzalez-Carranza, and J. A Roberts. 2003. "Plant Cell Wall Disassembly." In *Plant Cell Wall*, edited by J. K. C. Rose, vol. 8, pp. 264–324. Oxford: Blackwell Publishing Ltd.

Said, S., P. P. Agung, W. P.B. Putra, and E. M. Kaiin. 2020. "The Role of Biotechnology in Animal Production." *IOP Conference Series: Earth and Environmental Science* 492: 012035. doi:10.1088/1755-1315/492/1/012035.

Samuel, SM Sun. 2008. "Application of Agricultural Biotechnology to Improve Food Nutrition and Healthcare Products." *Asia Pacific Journal of Clinical Nutrition* 17(Suppl 1): 87–90. www.goldenrice.org.

Sánchez-León, Susana, Javier Gil-Humanes, Carmen V. Ozuna, María J. Giménez, Carolina Sousa, Daniel F. Voytas, and Francisco Barro. 2018. "Low-Gluten, Nontransgenic Wheat Engineered with CRISPR/Cas9." *Plant Biotechnology Journal* 16(4): 902–910. doi:10.1111/pbi.12837.

Sarkar, Surajit, and A. K. Mishra. 2001. "Bio-Preservation of Milk and Milk Products." *Indian Food Industry* 20(4): 74–77.

Sedeek, Khalid E.M., Ahmed Mahas, and Magdy Mahfouz. 2019. "Plant Genome Engineering for Targeted Improvement of Crop Traits." *Frontiers in Plant Science* 10: 114. doi:10.3389/fpls.2019.00114.

Sevillano, Laura, María T. Sanchez-Ballest, Felix Romojaro, and Francisco B. Flores. 2009. "Physiological, Hormonal and Molecular Mechanisms Regulating Chilling Injury in Horticultural Species. Postharvest Technologies Applied to Reduce Its Impact." *Journal of the Science of Food and Agriculture* 89(4): 555–563.doi:10.1002/jsfa.3468.

Sharma, Binny, and Asha Kumari. 2021. "Role of Post-Harvest Physiology in Evolution of Transgenic Crops." In *Plant Breeding: Current and Future Views*, edited by Ibrokhim Y. Abdurakhmonov, pp. 1–21. London: IntechOpen. doi:10.5772/intechopen.94694.

Sharma, Ram. Roshan., Dinesh Singh, and Rajbir Singh. 2009. "Biological Control of Postharvest Diseases of Fruits and Vegetables by Microbial Antagonists: A Review." *Biological Control* 50(3): 205–221. doi:10.1016/j.biocontrol.2009.05.001.

Singh, Akhilesh Kumar, Priti Pal, Uttam Kumar Sahoo, Laxuman Sharma, Brijesh Pandey, Anand Prakash, Prakash Kumar Sarangi, Piotr Prus, Raul Paşcalău, and Florin Imbrea. 2024. "Enhancing Crop Resilience: The Role of Plant Genetics, Transcription Factors, and Next-Generation Sequencing in Addressing Salt Stress." *International Journal of Molecular Sciences*. doi:10.3390/ijms252312537.

Singh, Dinesh, and Ram. Roshan. Sharma. 2018. "Postharvest Diseases of Fruits and Vegetables and Their Management." In *Postharvest Disinfection of Fruits and Vegetables*, edited by Mohammed Wasim Siddiqui, pp. 1–52. Amsterdam: Elsevier. doi:10.1016/B978-0-12-812698-1.00001-7.

Singh, Virenddra Bahadur. 2018. "Role of Biotechnology in Human Welfare." *Indian Journal of Scientific Research* 1: 71–74.

Sun, Samuel S.M., and Qiaoquan Liu. 2004. "Transgenic Approaches to Improve the Nutritional Quality of Plant Proteins." *Vitro Cellular & Developmental Biology–Plant 40*(2): 155–162. doi:10.1079/.IVP2003517

Tait, Joyce, and Guy Barker. 2011. "Global Food Security and the Governance of Modern Biotechnologies." *EMBO Reports* 12(8): 763–768. doi:10.1038/embor.2011.135.

Timmer, C Peter. 2003. "Presidential Lecture Biotechnology and Food Systems in Developing Countries 1 Biotechnology and Food Systems in Developing Countries." *Journal of Nutrition* 133(11): 3319–3322. https://www.econ.ox.ac.uk/.

Timperio, Anna Maria, Maria Giulia Egidi, and Lello Zolla. 2008. "Proteomics Applied on Plant Abiotic Stresses: Role of Heat Shock Proteins (HSP)." *Journal of Proteomics* 71(4): 391–411. doi:10.1016/j.jprot.2008.07.005.

Uncu, Ayse Ozgur, Sami Doganlar, and Anne Frary. 2013. "Biotechnology for Enhanced Nutritional Quality in Plants." *Critical Reviews in Plant Sciences* 32(5): 321–343. doi:10.1080/07352689.2013.781453.

Usigbe, Member Joy, Senorpe Asem-Hiablie, Daniel Dooyum Uyeh, Olayinka Iyiola, Tusan Park, and Rammohan Mallipeddi. 2024. "Enhancing Resilience in Agricultural Production Systems with AI-Based Technologies." *Environment, Development and Sustainability* 26: 21955–21983. doi:10.1007/s10668-023-03588-0.

Verhulst, Nele, Adrian Carrillo-García, Carina Moeller, Richard Trethowan, Ken D. Sayre, and Bram Govaerts. 2011. "Conservation Agriculture for Wheat-Based Cropping Systems under Gravity Irrigation: Increasing Resilience through Improved Soil Quality." *Plant and Soil* 340(1): 467–479. doi:10.1007/s11104-010-0620-y.

Voytas, Daniel F. 2013. "Plant Genome Engineering with Sequence-Specific Nucleases." *Annual Review of Plant Biology* 64: 327–350. doi:10.1146/annurev-arplant-042811-105552.

Wang, Wangxia, Basia Vinocur, Oded Shoseyov, and Arie Altman. 2004. "Role of Plant Heat-Shock Proteins and Molecular Chaperones in the Abiotic Stress Response." *Trends in Plant Science* 9(5): 244–252. doi:10.1016/j.tplants.2004.03.006.

Wang, Yanpeng, Xi Cheng, Qiwei Shan, Yi Zhang, Jinxing Liu, Caixia Gao, and Jin Long Qiu. 2014. "Simultaneous Editing of Three Homoeoalleles in Hexaploid Bread Wheat Confers Heritable Resistance to Powdery Mildew." *Nature Biotechnology* 32(9): 947–951. doi:10.1038/nbt.2969.

Wang, Yueju, Yuan Yeu Yau, Donna Perkins-Balding, and James G. Thomson. 2011. "Recombinase Technology: Applications and Possibilities." *Plant Cell Reports* 30(3): 267–285. doi:10.1007/s00299-010-0938-1.

Weber, Brittany N, Jon A Kobashigawa, and Michael M Givertz. 2017. "Evolving Areas in Heart Transplantation." *JACC: Heart Failure* 5(12): 869–878. doi:10.1016/j.jchf.2017.10.009.

Wu, Wei, Farooq Shah, and Bao-Luo Ma. 2022. "Understanding of Crop Lodging and Agronomic Strategies to Improve the Resilience of Rapeseed Production to Climate Change." *Crop and Environment* 1(2): 133–144. doi:10.1016/j.crope.2022.05.005.

Zhou, Jin R., and John W. Erdman. 1995. "Phytic Acid in Health and Disease." *Critical Reviews in Food Science and Nutrition* 35(6): 495–508. doi:10.1080/10408399509527712.

7 Prospects of Integrated Farming System

Tank Sinderpal Kabalsingh, Mandeep Kaur, Prabha, and Archana Saini

7.1 INTRODUCTION

Integrated farming system (IFS) forms a comprehensive agricultural model that aims to improve environmental sustainability, resource efficiency, and productivity as it brings together several farm businesses such as crops, livestock, aquaculture, and agroforestry (Atapattu et al. 2025). This strategy supports farmers' resilience and diversification of revenue sources, which is in line with the global sustainable development goals (SDGs, Bhargavi and Behera 2020). IFS adoption has drawbacks despite its advantages, including the need for institutional support, market connections for a variety of farm products, and knowledge and skill requirements. The upfront costs and management complexity are frequent challenges for smallholder farmers (Mandal et al. 2017). Research IFS models in India and other nations show notable rises in farm output and income, sometimes doubling returns over monoculture systems (Bhargavi and Behera 2020). Promising in their ability to balance environmental and financial objectives, community-based IFS models tailored for local conditions have IFS's future lies in its integration with modern technologies and extension services to increase scalability and adoption (Atapattu et al. 2025). To reach sustainable development by means of region-specific IFS models, ongoing research and policy support are absolutely essential. Recent studies show that by increasing resource efficiency and resilience, IFS is absolutely vital in helping agriculture to adapt to climate change and sustainability challenges (Jagustovic et al. 2021). Innovations in farming systems are crucial for enhancing productivity by reducing environmental footprints (Singh et al. 2020).

The principal goal of IFS is to optimize the utilization of all available resources through in situ recycling and diversification, thereby improving livelihood security, promoting environmental stewardship, and ensuring economic viability for farmers, particularly those with marginal and small landholdings (Behera and France 2016). IFS is not merely a combination of enterprises, but a dynamic system that ensures the complementary and supplementary role of each component to generate sustainable farm outputs with minimum external inputs (Singh and Kumawat 2023). This integration aids in risk mitigation, nutrient recycling, and sustainable farm output maximization.

Although formalized scientific studies have increased recently with the focus on sustainability, integrated farming practices have historically been incorporated into

DOI: 10.1201/9781003452805-7

TABLE 7.1
Comparative Analysis of Monoculture Versus Integrated Farming System

Aspect	Monoculture System	Integrated Farming System	Benefits of IFS
Crop diversity	Low	High	Reduces risk and improves resilience
Resource use efficiency	Low	High	Optimizes water, nutrients, and energy use
Environmental impact	High (pollution, degradation)	Low (sustainable practices)	Reduces ecological footprint
Income stability	Unstable	Stable and diversified	Multiple revenue streams
Employment generation	Seasonal	Year-round	Engages labor across enterprises

traditional agricultural systems in many countries (Mandal et al. 2017). As their importance in preserving resources and enhancing rural livelihoods has become more widely acknowledged, IFS models have changed. Historically, integrated and varied approaches were used in traditional agricultural practices in many civilizations, especially in the Indian subcontinent. Prior to the Green Revolution, farmers in India engaged in mixed farming and heavily relied on traditional knowledge systems. For instance, trees supplied food, fodder, and fuel, while livestock provided organic manure and draft power. However, agricultural strategies underwent a dramatic shift toward monoculture and high external input-based farming systems with the onset of the Green Revolution in the 1960s. This paradigm shift led to a number of unintended consequences, such as soil degradation, nutrient imbalance, groundwater resource depletion, biodiversity loss, and increased production costs, even though it greatly increased the production of food grains (NAAS 2013; FAO 2018). Research organizations like the Indian Council of Agricultural Research (ICAR), which started advocating for region-specific IFS frameworks in the early 2000s, have shown renewed interest in IFS models as a result of smallholder farmers' growing susceptibility to environmental stressors, market volatility, and climate change (Table 7.1).

IFS provide a viable alternative to farmers' current problems with climate change, resource depletion, and economic vulnerability by strengthening agroecosystem resilience and encouraging a variety of revenue streams (Bhargavi and Behera 2020). Adoption of it can increase food security and lessen reliance on outside inputs (Atapattu et al. 2025). The urgent need to make farming economically feasible, environmentally sustainable and socially inclusive highlights the importance of IFS in today's agricultural discourse. Despite their initial increases in productivity, traditional monoculture systems are no longer feasible because of rising input costs, diminishing returns, and growing ecological footprints. By encouraging resource efficiency, waste reduction, income diversification, and environmental preservation, IFS provides a route toward sustainable intensification (Soni et al. 2014).

Furthermore, in light of market volatility and climate change, IFS's capacity to reduce risk through diversification is especially important. IFS is a strategic tool for attaining several SDGs, such as food security (SDG 2), poverty alleviation (SDG 1), climate action (SDG 13), and responsible consumption and production (SDG 12), by conforming to the agroecology and circular economy principles (FAO 2018). When several farming businesses are integrated into IFS, smallholder livelihoods are improved by stabilizing farm economies and diversifying sources of income, particularly in light of changing climate conditions. Additionally, these systems have a major impact on ecosystem services and soil health.

7.2 COMPONENTS OF INTEGRATED FARMING SYSTEM

To maximize nutrient flows and farm productivity, different elements—crop production, livestock, aquaculture, agroforestry, and other ancillary businesses—work together. By enhancing biodiversity, nutrient recycling, and income diversification, IFS promotes ecological, social, and economic sustainability (Atapattu et al. 2025). By limiting chemical inputs, the system lessens its effects on the environment and promotes social inclusion by offering a variety of livelihood options (Bhargavi and Behera 2020). By enhancing microclimates and carbon sequestration, agroforestry and other diverse elements of IFS contribute to climate resilience. Aquaculture, crops, and livestock all work together to improve farm sustainability and nutrient cycling (Singh et al. 2020).

7.2.1 Crop Production

Crop production forms the bedrock of any IFS, providing not only food for the farm family and market but also essential raw materials for other enterprises such as fodder for livestock, feed for fish, and substrates for mushroom cultivation (Meena et al. 2024). In IFS, the selection of crops is based on agro-climatic conditions, market demand, and the synergy with other farm enterprises. The practice emphasizes multitiered cropping systems, including intercropping, crop rotation, and relay cropping, which are crucial for maintaining soil fertility, breaking pest and disease cycles, and enhancing yield stability (Singh and Kumawat 2023). Additionally, organic nutrient management practices such as the application of compost, green manures, biofertilizers, and vermicompost are strongly advocated to maintain soil health and reduce the dependency on chemical fertilizers (Behera and France 2016). The integration of pulses and legumes into the cropping system further improves nitrogen fixation, leading to enhanced productivity and sustainability.

7.2.2 Livestock Integration

The contribution of livestock to income, nutrition, soil fertility, and energy security makes it a crucial component of the IFS. Pigs, sheep, poultry, goats, cattle, and buffaloes are frequently included based on local tastes and the availability of resources. By using animal waste as organic manure, the symbiotic interaction between livestock and crop operations guarantees nutrient recycling; this lowers fertilizer costs and improves soil fertility and structure (Gupta et al. 2020). Livestock contributes to the stability of agricultural revenue by generating a steady cash flow through the sale of milk, meat, eggs, and wool in addition to manure. Since women do the majority of animal husbandry work in many rural areas, animals play a crucial socioeconomic role by creating jobs and empowering them (Soni et al. 2014).

7.2.3 Aquaculture

The integration of aquaculture into farming systems, particularly in water-rich and lowland areas, is increasingly gaining traction due to its potential to provide high-value protein, income diversification, and optimal water resource utilization. Integrated rice–fish farming, for instance, exemplifies a symbiotic system wherein fish control pests and weeds in paddy fields while benefiting from the nutrients present in flooded conditions. Fish ponds can also be utilized for irrigation, while pond silt, rich in nutrients, can be periodically applied to crop fields. Moreover, integrating fish with duck or poultry rearing further enhances nutrient cycling and reduces feeding costs, leading to improved resource use efficiency. Such integrated aquaculture models contribute significantly to food and nutritional security, especially for land-limited and resource-poor households.

7.2.4 Agroforestry

Agroforestry, which is the deliberate planting of trees and shrubs in agricultural landscapes, fulfils a number of ecological and financial purposes in IFS. It supports livelihoods and biodiversity protection by offering shade, windbreaks, fuelwood, feed, timber, and non-timber forest products. Because of their ability to fix nitrogen and their many uses, trees including *Gliricidia sepium*, *Leucaena leucocephala*, *Moringa oleifera*, and *Azadirachta indica* are frequently included (Ayyam et al. 2019). Agroforestry systems decrease soil erosion, stabilize microclimates, and improve carbon sequestration. Perennial woody components also serve as a risk buffer and provide long-term financial benefits when incorporated into the farming system. Agroforestry helps retain moisture and provide essential ecosystem services that are necessary for long-term sustainability in dry and semiarid regions.

7.2.5 Apiculture, Sericulture, and Mushroom Cultivation

In Integrated farming, these micro businesses play a vital role in creating jobs and diversifying revenue streams, although frequently being overlooked. **Apiculture (beekeeping)** by providing pollination services, increases crop yield and generates revenue from beeswax, honey, and other hive products. A mutually advantageous system is produced when apiculture is combined with floral crops and fruit orchards (TNAU 2018). **Sericulture, or silkworm farming,** is mostly performed in Karnataka, Assam, and West Bengal. It supports agro-based enterprises by using mulberry crops and making money from the sale of raw silk. **Mushroom cultivation** uses agricultural waste materials like sawdust and straw as substrates, turning trash into a high-protein meal. Shiitake, button, and oyster mushrooms are perfect for smallholders since they require little area and resources to grow. According to Soni et al. (2014), these businesses are low-investment, high-return endeavors that empower rural women and youth through microenterprise development and offer nutritional supplements.

7.2.6 Waste Recycling and Bioenergy

IFS's capacity to efficiently recycle organic leftovers and turn waste into wealth is one of its distinguishing characteristics. Bio-compost, vermicompost, or biogas is produced from biodegradable materials such as animal manure, agricultural leftovers, and kitchen trash. In instance, biogas technology has two advantages: it produces nutrient-rich slurry for fields and clean cooking fuel (Singh and Kumawat 2023). The energy efficiency and environmental impact of agricultural systems are further improved by the use of renewable energy sources like biomass gasifiers, wind turbines, and solar-powered irrigation. Reducing greenhouse gas emissions, minimizing reliance on fossil fuels, and attaining closed-loop, zero-waste agriculture are all made possible by these waste management and bioenergy systems. Over time, these technologies increase farms' economic efficiency and support rural development that is sustainable. By decreasing chemical inputs and increasing soil organic matter, IFS promotes ecological balance and lowers greenhouse gas emissions, hence promoting environmental sustainability. Research shows that when IFS is broadly implemented, agricultural revenue rises and poverty decreases (Raihan 2024).

7.3 BENEFITS OF INTEGRATED FARMING SYSTEM

IFS offers a multitude of benefits, especially for small and marginal farmers who rely on agriculture for their livelihood. One of the foremost advantages is enhanced resource utilization, as it promotes the recycling of farm waste to ensure optimal use of available resources. For instance, crop residues serve as fodder, while animal manure is applied as organic fertilizer (Fatima et al. 2023). Dar et al. (2018) demonstrated a model where rice, poultry, and mushroom cultivation were integrated to create a closed-loop system, using spent mushroom beds as cattle feed and rice straw as a mushroom substrate, while poultry droppings fertilized rice fields. Similarly, Mahapatra and Behera (2011) highlighted an IFS model in Odisha that incorporated vegetable farming, duck rearing, and aquaculture, where duck droppings enhanced plankton growth in ponds for fish feed, and nutrient-rich pond silt was recycled into vegetable plots. This integration significantly boosts farm income and employment opportunities by diversifying revenue sources, reducing dependence on a single commodity, and providing steady cash flow from livestock and poultry sales, even during crop off-seasons (Singh and Kumawat 2023). Das et al. (2019) observed that IFS increased annual employment to 280 man-days compared to 120–140 in conventional systems, thereby reducing rural underemployment. Additionally, IFS fosters sustainability and environmental protection by increasing on-farm biodiversity, minimizing chemical inputs through practices such as integrated pest management, and enhancing ecological resilience (Singh and Kumawat 2023). It ensures food and nutritional security by maintaining a steady supply of diverse food items grains, vegetables, fruits, milk, meat, and eggs (Dar et al. 2018; Mahapatra and Behera 2011). Moreover, by diversifying income sources, IFS mitigates risks associated with climatic and market shocks. For example, during erratic monsoons in Eastern India, income from vegetables, goat rearing, and fish farming offset losses from reduced paddy yields (Das et al. 2019). The system also promotes soil health through crop rotation and organic inputs, which improve soil organic carbon and microbial activity (Mahapatra and Behera 2011). Finally, IFS enhances energy efficiency by incorporating renewable resources like biogas and animal power, thus reducing dependence on fossil fuels. In Uttar Pradesh, biogas generated from cattle manure was used for cooking and lighting, and the nutrient-rich slurry was reapplied to fields, while bullocks replaced diesel-powered machinery for tilling (Das et al. 2019), contributing to a more energy-efficient and sustainable farming system.

7.4 CHALLENGES IN ADOPTION OF INTEGRATED FARMING SYSTEM

Despite its numerous benefits, the widespread adoption of IFS faces several significant challenges, particularly for small and marginal farmers. One of the primary barriers is the lack of technological awareness and institutional support. Smallholders often do not have access to IFS-related technologies or the technical expertise needed to design and manage the interrelated components effectively (Fatima et al. 2023). This knowledge gap, coupled with limited extension services and inadequate training opportunities, hampers the practical implementation of IFS, leading to suboptimal resource use and reduced productivity. Moreover, institutional frameworks are often geared toward single-crop systems, failing to provide the integrated guidance required for diversified farming (Singh and Kumawat 2023). Financial constraints also pose a significant challenge; the initial investment needed for setting up infrastructure for livestock, fisheries, or biogas units is often beyond the means of small-scale farmers, who also struggle to access credit and subsidies (Das et al. 2019). Labor and management complexities further hinder adoption, as IFS demands intensive labor and efficient coordination among various farm components, which can be overwhelming for households with limited workforce availability (Dar et al. 2018). Additionally, insufficient market access and infrastructure for perishable IFS products like milk, vegetables, and fish undermine profitability, as fragmented and inefficient supply chains prevent timely and fair-market transactions (Mahapatra and Behera 2011). Land fragmentation is another critical issue, especially in developing countries where small, scattered plots limit the feasibility of integrating diverse enterprises like aquaculture and agroforestry (Das et al. 2019). These fragmented holdings often lack the contiguous space necessary for setting up functional IFS units, complicating land use planning and reducing the benefits of diversification. Sociocultural factors also play a role; traditional beliefs, social norms, and gender roles can influence decision-making and hinder the integration of components like animal husbandry or aquaculture, particularly in conservative or resource-constrained communities (Singh and Kumawat 2023). Together, these challenges highlight the need for targeted policy reforms, inclusive extension strategies, and infrastructural support to facilitate broader adoption of IFS.

7.5 STRATEGIES FOR OVERCOMING CHALLENGES

To overcome the challenges associated with the adoption of IFS, several strategic interventions are essential. Strengthening capacity building and extension services is critical; empowering farmers through training programs and participatory learning methods, such as Farmer Field Schools (FFS), has been shown to significantly enhance the understanding and adoption of IFS practices (Dar et al. 2018). Financial support is equally vital provision of low-interest loans, targeted subsidies, and access to microfinance can help smallholders establish the necessary infrastructure. Government schemes such as the Rashtriya Krishi Vikas Yojana (RKVY), National Mission for Sustainable Agriculture (NMSA), and Paramparagat Krishi Vikas Yojana (PKVY) offer essential financial and technical assistance to promote IFS adoption. Research and development must also be prioritized; as Fatima et al. (2023) suggest, region-specific IFS models should be developed based on local agro-climatic conditions, resource availability, and farmer preferences. Alongside this, integrated policy reforms are necessary to facilitate interdepartmental coordination among agriculture, animal husbandry, fisheries, and rural development, thereby promoting sustainable and inclusive farming practices. Enhancing market access through investments in rural infrastructure, cold storage, and value addition can significantly boost profitability. Associating IFS produce with niche markets such as organic and farm-fresh segments, and promoting cooperative marketing, particularly through Farmer Producer Organizations (FPOs), allows smallholders to negotiate better prices and tap into larger markets (Kumar et al. 2015). Community-based approaches, including FPOs, cooperatives, and self-help groups (SHGs), can address individual constraints and enhance collective resilience. These institutions not only facilitate knowledge dissemination and input access but also improve market bargaining power, thereby enabling efficient and scalable adoption of IFS models across diverse regions.

7.6 CASE STUDIES AND SUCCESS STORIES

IFS in India have been extensively promoted to mitigate the problems of small and marginal farmers, including land fragmentation, poverty, and environmental degradation. An example of this is the crop–livestock–fishery IFS in Sundarbans of West Bengal, where rice is cultivated along with fish rearing and duck rearing. This model effectively recycles resources like manure and feed, leading to higher productivity and income along with improved household nutritional security (Sati and Vangchhia 2017). Tamil Nadu Agricultural University (TNAU) in Tamil Nadu has created IFS models for dryland farmers that include goat rearing, millets, horticulture, and agroforestry. These systems enhance soil fertility through recycling and provide drought resilience. Likewise, indigenous IFS practices in north-eastern states such as Nagaland and Mizoram combine piggery, poultry, bamboo cultivation, and jhum (shifting) cultivation in a sustainable pattern. These area-specific models prove IFS versatility to varied agro-climatic zones (Shyam et al. 2023). Institutional support has been essential in disseminating IFS. Krishi Vigyan Kendras (KVKs) throughout India have played a vital role in exhibiting viable IFS models and training farmers in integrated methods.

IFs in Southeast Asia have its roots in traditional agriculture but has been increasingly valued because of growing anxiety over food security and environmental sustainability. A typical example in Vietnam is the VAC system (Vuon-Ao-Chuong), which is a classical homestead-based IFS model. It combines vegetable and fruit cultivation with aquaculture and livestock raising, providing maximum land use efficiency. Livestock organic wastes are also transformed into compost and fish feed, eliminating the requirement for external inputs. In Java, Indonesia, the rice–fish–duck model integrates rice cultivation with fish rearing and duck keeping. It has proven to minimize pest infestation and chemical fertilizer use, thus improving ecological harmony (Setboonsarng 2006). Integrated rice–duck farming has been promoted in the Philippines under the guidance of the Department of Agriculture and NGOs. Ducks provide natural control of pests and improve soil fertility, thereby raising both paddy and farm incomes (Liu et al. 2020). These Southeast Asian IFS models enjoy good community participation, indigenous knowledge, and active government support. Financial and technical assistance from local authorities and cooperatives has further accelerated adoption.

A comparative study between Indian and Southeast Asian IFS models shows both similarities and region-specific differences. Both regions share the same overall principles of resource recycling and integration of several enterprises to enhance sustainability, food security, and income. Yet, Southeast Asian models like Vietnam's VAC model are generally more local and homestead-based, whereas Indian models cover wider rural areas and often involve agroforestry and rainfed farm management practices (Shyam et al. 2023; Phong et al. 2010). Support from policies is also variable. Southeast Asia, and Vietnam and the Philippines in particular, has top-down support with well-structured extension services. Indian support systems, while existing, are frequently fragmented and less available to distant populations (Singh et al. 2018). Southeast Asian models contain more indigenous knowledge and cultural practices deeper within them, while traditional wisdom in India is integrated to varying extents by area. Regardless of differences, IFS improves resilience to climatic shocks, maximizes the use of available resources, and enhances profitability in each region. These revelations emphasize the need to adapt IFS to agro-ecological and socioeconomic conditions of the local setting for sustainable success. Comparative analysis of IFS models in India and Southeast Asia is presented in Table 7.2.

7.7 ROLE OF GOVERNMENT AND INSTITUTIONS

IFS is a sustainable farming system that integrates different farm enterprises crops, livestock, aquaculture, horticulture, agroforestry, and beekeeping to increase productivity, diversify income streams, and improve climate resilience. In India, government policies, institutional setup, and research institutions have also been at the forefront of promoting and upscaling IFS models, particularly for small and marginal farmers.

The Government of India has launched various policy platforms and financial programs to promote IFS. The NMSA encourages location-specific adoption of IFS through modules such as Rainfed Area Development and Soil Health Management (Ministry of Agriculture and Farmers Welfare 2020). The ICAR coordinates the All India Coordinated Research Project on Integrated Farming Systems (AICRP-IFS) to formulate region-specific, resource-constrained models (ICAR 2021).

TABLE 7.2
Comparative Analysis of IFS Models in India and Southeast Asia

Aspect	India	Southeast Asia
Common IFS models	Crop–livestock–fishery, agroforestry, horticulture–poultry–goat combinations	VAC system (Vietnam), rice–fish–duck (Indonesia, Philippines)
Integration focus	Broad-scale, often on rural or semi-rural lands	Homestead-based and small-scale intensification
Resource recycling	Use of crop residues, livestock manure, biogas for energy	Household waste used in fish ponds, livestock waste used in gardens
Policy and institutional support	National schemes (e.g., NMSA, RKVY), KVKs, ICAR initiatives	Strong local government, cooperative, and community involvement
Use of indigenous knowledge	Present, but variable across regions	Deeply rooted and integrated into farming systems
Environmental benefits	Reduced chemical inputs, improved soil health, water use efficiency	Natural pest control (e.g., ducks), organic nutrient cycling
Income enhancement	25%–50% increase over monoculture	Significant boost due to year-round diversified produce
Major constraints	Land fragmentation, limited awareness, inconsistent support	Market linkage issues, urban encroachment on farmland
Success drivers	Demonstrations by KVKs, local adaptation, farmer cooperatives	Tradition, compact system efficiency, government support
Examples	Rice–fish–duck system in West Bengal, millet–goat–agroforestry in Tamil Nadu	VAC system in Vietnam, rice–fish–duck in Philippines and Indonesia

Other programs like RKVY and PKVY promote organic and diversified agriculture. More recent programs like the Digital Agriculture Mission (2024) and National Mission on Natural Farming (NMNF) also bring modern technology and traditional IFS practices more together. Initiatives such as the Namo Drone Didi Scheme empower women in precision farming, while state-level programs like Tamil Nadu's IFS Units and the "One Village One Product" scheme integrate IFS into local development plans (The Hindu 2024).

Extension activities are needed to fill the gap between research and farmers. In India, central governments, state governments, agricultural universities, and KVKs provide training, demonstrations, and exposure visits suited to IFS. KVKs help in the capacity building, technology transfer, and place-specific model implementation (ICAR 2021).

Under the NMSA, specialized funds finance climate-resilient practice training as well as IFS components that are appropriate for local conditions (Ministry of Agriculture and Farmers Welfare 2020). The Sub Mission on Agricultural Extension (SMAE) under the National Mission on Agricultural Extension and

Technology (NMAET) also strengthens advisory services through Agri-Clinics and Agri-Business Centres (ACABCs).

Tamil Nadu and other states have FFS and demonstration plots as part of their IFS programs focusing on integrated nutrient management and organic inputs (The Hindu 2024). Electronic platforms such as eNAM, Kisan Call Centres, and mobile applications (e.g., Kisan Suvidha) provide up-to-date information. The Digital Agriculture Mission enables the creation of personalized training and extension services through digital platforms.

Research and development form the basis of the development of IFS. The AICRP-IFS by ICAR compares IFS models over various agro-climatic zones, including enterprises such as crops, livestock, horticulture, fisheries, and agroforestry (ICAR 2021). The models are designed to optimize land use, resource use efficiency, and income and ensure ecological balance. The areas of research priorities are nutrient cycling, productivity evaluation, economic sustainability, and environmental analysis. Research indicates diversified IFS models perform better than monoculture systems in productivity and net income, especially for small farmers (ICAR 2020). Composting, water harvesting, and organic inputs are also areas of prime importance. State agricultural universities play their role in research and development by tailoring models, conducting long-term experiments, and liaising with KVKs for field demonstrations and farmer training. Digital agriculture technologies like AgriStack and decision support systems improve IFS accuracy and flexibility. International partnerships with food and agriculture organization (FAO) and international maize and wheat improvement center enhance system diversification and climate-smart innovation.

7.8 FUTURE PROSPECTS AND INNOVATIONS

With accelerating climate variability, resource degradation, and food security concerns, IFS are gaining prominence as a visionary and adaptive solution. New innovations and supportive policy frameworks are transforming IFS into resilient, productive, and market-driven models capable of addressing multidimensional rural challenges.

India's agricultural sector, particularly for smallholders, is increasingly vulnerable to climate extremes like erratic rainfall, droughts, floods, and rising temperatures. In response, climate-resilient IFS models have been developed under the ICAR's National Innovations on Climate Resilient Agriculture (NICRA) program. These models integrate enterprises such as crops, livestock, fisheries, poultry, agroforestry, and beekeeping, tailored to the specific agroecological conditions of each region (Reddy et al. 2018). For instance, flood-prone Assam benefits from rice–fish–duck farming, while drought-prone Bundelkhand emphasizes goat rearing, poultry, and drought-tolerant crops.

Key climate-adaptive features in these IFS models are the following:

- Year-round production using water harvesting, check dams, and micro-irrigation.
- Soil fertility restoration through composting, green manuring, and carbon-sequestering practices.

- Crop diversification and agroforestry for spreading risk and improving resilience.
- Livestock and aquaculture integration to buffer income in the event of crop failure.

According to ICAR-NICRA (2021), these models can increase net returns by 2–3 times compared to monocropping, while also reducing greenhouse gas emissions due to lower synthetic input use and improved soil carbon retention. Complementary government schemes such as the NMSA and State Action Plans on Climate Change (SAPCCs) support infrastructure for water conservation, renewable energy, and community seed banks.

Technology is reshaping IFS into smart, data-driven systems. Precision agriculture tools like drones, GPS-enabled machinery, and soil sensors, enable site-specific input management, optimize resource use, and minimize environmental footprints. Programs like Namo Drone Didi are empowering women-led SHGs to operate drones for spraying and farm monitoring, enhancing IFS productivity (Krishak Jagat 2024). The Digital Agriculture Mission and AgriStack platforms are integrating satellite imagery, weather data, soil health indicators, and crop analytics to provide real-time, personalized farm advice. These technologies facilitate synchronized management of diverse enterprises like crops, livestock, and aquaculture, adapting dynamically to climatic and market changes (Atapattu et al. 2024; Sharma et al. 2022).

Innovations in in bio-inputs such as microbial biofertilizers, biopesticides, and vermicomposting spur organic farming and soil health. Solar irrigation pumps, smart irrigation, and automated feeders for livestock promote energy and resource savings. For aquaculture-based IFS, recirculating aquaculture systems and automated aerators help improve fish output in small spaces. Greenhouse and polyhouse technologies coupled with IFS allow cultivation of high-value crops throughout the year, even in climatically sensitive areas. R&D centers such as ICAR and state agricultural universities are also testing Internet of Things (IoT) and artificial intelligence (AI)-based applications in IFS. Some of these include pest detection systems, remote monitoring of animal health, and AI-based yield forecasting. Major technological interventions in IFS are given in Table 7.3.

Further, a critical driver for IFS viability is effective market linkage and value addition. Integrated farms offer diverse produce—milk, eggs, fish, fruits, vegetables,

TABLE 7.3
Key Technological Interventions in IFS

Technology	Application Area	Benefit
Drones	Crop spraying, monitoring	Precision input use, reduced labor
Soil sensors	Soil health assessment	Improved nutrient management
AgriStack platform	Digital decision support	Real-time data and advisory
Recirculating aquaculture systems (fish farming)	Aquaculture	Space-efficient and water-saving
IoT devices	Livestock, environment	Automated monitoring and alerts

TABLE 7.4
Value Addition Opportunities in IFS

Enterprise	Value-Added Product	Market Potential
Dairy	Paneer, ghee, flavored milk	Urban retail and cooperatives
Horticulture	Pickles, juices, dehydrated veg	Export and health food market
Fishery	Smoked fish, fish pickles	Local and gourmet markets
Beekeeping	Herbal honey, beeswax candles	Organic and cosmetic industries

honey, herbs—creating opportunities for both rural and urban market access. The e-NAM (electronic National Agricultural Market) platform enables transparent price discovery and national market integration for IFS produce (Ministry of Agriculture and Farmers Welfare 2022).

FPOs play a pivotal role in collective marketing, bargaining for better prices, and facilitating aggregation of niche products like organic vegetables, dairy, and herbal goods. Value addition like milk to paneer or ghee, fruits to jam, herbs to teas enhances shelf life and profitability. This is supported by training and handholding through KVKs, Rural Self Employment Training Institutes (RSETIs), and FPOs, which provide skill development in food processing, branding, and packaging.

Government schemes like PM Formalisation of Micro Food Processing Enterprises (PM-FME) (MOFPI 2023) offer credit and subsidy support to boost rural agro-processing. Cold chains, food parks, and modernized rural haats, enabled through the Agri Infrastructure Fund, reduce postharvest losses and improve logistics. Innovative initiatives like "One Village One Product" (Tamil Nadu), localized branding, and use of mobile apps and virtual marketplaces further strengthen rural-urban value chains. Direct-to-consumer models enhance farmer margins by reducing middlemen, creating a more equitable agri-value ecosystem. Value addition opportunities in IFS are depicted in Table 7.4.

7.9 CONCLUSION

IFS represent a transformative paradigm for sustainable agriculture, particularly for small and marginal farmers navigating ecological degradation, economic vulnerability, and climate uncertainty. By synergizing diverse farm components crops, livestock, aquaculture, agroforestry, and bioenergy, IFS creates closed-loop systems that enhance biodiversity, recycle nutrients, reduce input dependency, and diversify income streams, ultimately leading to greater resilience and sustainability. Unlike conventional monoculture systems that often exhaust natural resources and increase climate sensitivity, IFS promotes ecological intensification—producing more from available land while preserving environmental quality. System-level efficiency is achieved by aligning outputs and by-products of one enterprise (e.g., livestock manure, crop residues) as inputs for another, thus minimizing waste and reducing external input costs. Comparative studies of IFS in India and Southeast Asia reveal their context-specific adaptability and scalability. In Vietnam, for example, rice–fish–vegetable systems have increased per-hectare income and improved household nutrition. Similarly, India's regionally adapted IFS models such as rice–fish–duck in Assam or goat–horticulture in arid Rajasthan demonstrate that when aligned with local agroecological and sociocultural contexts, IFS becomes both economically viable and socially acceptable. However, transitioning from fragmented or monoculture systems to integrated models is not without challenges. Farmers often face constraints related to initial investment, access to credit, land size fragmentation, technical know-how, and inadequate institutional support. Additionally, labor intensiveness of diversified systems and risk aversion among farmers can slow adoption. To address these barriers, robust institutional frameworks are essential providing financial incentives, crop-livestock insurance, climate services, and extension outreach. Capacity building through FFS, KVKs, and community-based learning platforms can facilitate knowledge exchange, while rural credit schemes and subsidies can ease the financial transition to integrated systems. The future success of IFS hinges on its mainstreaming into national and subnational agricultural development strategies. This includes embedding IFS into programs like the NMSA, PM-KISAN, and state agro-climatic plans. Digital agriculture tools like IoT sensors, mobile apps, geospatial data, AI-driven advisory services are also revolutionizing farm-level decision-making, allowing real-time optimization of complex IFS components. Furthermore, community-driven approaches such as FPOs, SHGs, and women-led cooperatives play a critical role in scaling IFS, enabling collective procurement, processing, branding, and marketing. Investments in rural infrastructure cold chains, agro-processing units, rural markets, and renewable energy are vital for value addition and economic viability. As agriculture confronts the dual challenge of feeding a growing population and mitigating climate risks, IFS emerges as a resilient, inclusive, and regenerative solution. Its integrated nature makes it ideal for addressing food and nutritional security, livelihood diversification, resource conservation, and climate adaptation, making IFS a cornerstone for future-ready agriculture and empowered rural economies.

REFERENCES

Atapattu, A.J., Nuwarapaksha, T.D., Udumann, S.S., Dissanayaka, N.S. 2025. Integrated farming systems: A holistic approach to sustainable agriculture. In: Babu, S., Kumar, R., Singh, P., Sharma, A. (Eds.), *Agricultural Diversification for Sustainable Food Production*. Springer, New York, pp. 89–110.

Atapattu, A.J., Ranasinghe, C.S., Nuwarapaksha, T.D., Udumann, S.S., Dissanayaka, N.S. 2024. Sustainable agriculture and sustainable development goals (SDGs). In: Garwi, J., Masengu, R., Chiwaridzo, O. (Eds.), *Emerging Technologies and Marketing Strategies for Sustainable Agriculture*. IGI Global, Hershey, pp. 1–27

Ayyam, V., Palanivel, S., Chandrakasan, S. 2019. Agroforestry for livelihood and biodiversity conservation. In: Ayyam, V., Palanivel, S., Chandrakasan, S. (Eds.), *Coastal Ecosystems of the Tropics Adaptive Management*. Springer, Singapore, pp. 363–389.

Behera, U.K., France, J. 2016. Integrated farming systems and the livelihood security of small and marginal farmers in India and other developing countries. *Adv. Agron.* 138: 235–282.

Bhargavi, B., Behera, U. 2020. Securing the livelihood of small and marginal farmers by diversifying farming systems. *Curr. Sci.* 119(5): 854–860.

Dar, N.A., Lone, B.A., Alaie, B.A., Dar, Z.A., Gulzafar, Bahar, F.A., Haque, S.A., Singh, K.N. 2018. Integrated farming systems for sustainable agriculture. In: Sengar, R., Singh, A. (Eds.), *Eco-Friendly Agro-Biological Techniques for Enhancing Crop Productivity*. Springer, Singapore, pp. 111–127.

Das, A., Layek, J., Ramkrushna, G., Subhash, B., Devi, M.T., Dey, U., Suting, D., Yadav, G.S., Lyngdoh, D.B.D., Prakash, N. 2019. Integrated organic farming system: An innovative approach for enhancing productivity and income of farmers in North Eastern Hill Region of India. *Indian J. Agric. Sci.* 89(8): 1267–1272.

FAO. 2018. *The Future of Food and Agriculture: Alternative Pathways to 2050*. Food and Agriculture Organization of the United Nations. https://www.fao.org/3/I8429EN/i8429en.pdf.

Fatima, A., Singh, V.K., Subhash, B., Singh, R.K., Upadhyay, P.K., Rathore, S.S., Kumar, B., Hasanain, M., Parween, H. 2023. Food production potential and environmental sustainability of different farming system models in northwest India. *Front. Sustain. Food Syst.* 7: 959464.

Gupta, G., Palsaniya, D.R., Manoj, S., Upadhyay, D., Suman, M. and Chand, K. 2020. Livestock based integrated farming system. *Indian Farm.* 70(6). doi:10.59797/ija.v54i2.4790.

ICAR. 2020. *Annual Report 2019–2020*. Indian Council of Agricultural Research. https://icar.org.in/.

ICAR. 2021. *Annual Report 2020–2021*. Indian Council of Agricultural Research. https://icar.org.in/.

ICAR-NICRA. 2021. *Annual Progress Report 2020–2021*. https://www.atarikolkata.org/wp-content/uploads/2022/03/NICRA-Annual-Report-2020-21-compressed.pdf

Jagustovic, R., Papachristos, G., Zougmoré, R.B., Kotir, J.H., Kessler, A., Ouédraogo, M., Dittmer, K.M. 2021. Better before worse trajectories in food systems? An investigation of synergies and trade-offs through climate-smart agriculture and system dynamics. *Agric. Syst.* 190: 103131.

Krishak Jagat. 2024. *New Schemes for Farmers*. https://www.en.krishakjagat.org/.

Kumar, A., Singh, R.K., Singh, P. 2015. Role of farmer producer organizations in improving market access and income of farmers. *Ind. J. Agric. Eco.* 70(3): 357–368.

Liu, S., Li, X., Chen, D., Duan, Y., Ji, H., Zhang, L., Chai, Q., Hu, X. 2020. Understanding land use/land cover dynamics and impacts of human activities in the Mekong Delta over the last 40 years. *Global Ecol. Conserv.* 22: e00991.

Mahapatra, I.C., Behera, U.K. 2011. Integrated farming systems for livelihood security of small and marginal farmers in India. *Ind. J Agron.* 56(4): 251–260.

Mandal, U., Panda, R.K., Mishra, P.K., Alam, N.M., Kar, G. 2017. Conceptualization of community-based integrated farming system model design with multi-objective optimization management. *Curr. Sci.* 112(11): 2232–2242.

Meena L.R., Kochewad S.A., Kumar D., Malik Suresh, Meena S.R. 2024. Development of sustainable integrated farming systems for small and marginal farmers and ecosystem services: A comprehensive review. *Agricul. Sci. Dig.* 44(3): 391–397. https://doi.org/10.18805/ag.D-5961.

Ministry of Agriculture and Farmers Welfare. 2020. *NMSA Guidelines*. https://nmsa.dac.gov.in/.

Ministry of Agriculture and Farmers Welfare. 2022. *e-NAM: Electronic National Agriculture Market*. https://www.enam.gov.in/.

MOFPI. 2023. *PM-FME Scheme Guidelines*. https://mofpi.gov.in/

NAAS. 2013. *Integrated Farming Systems: Issues and Strategies* (Policy Paper 66). National Academy of Agricultural Sciences, New Delhi.

Phong, L.T., van Dam, A.A., Udo, H.M.J., van Mensvoort, M.E.F., Tri, L.Q., Steenstra, F.A., van der Zijpp, A.J. 2010. An agro-ecological evaluation of aquaculture integration into farming systems of the Mekong Delta. *Agricult. Ecosys. Environ.* 138: 232–241. doi:10.1016/j.agee.2010.05.004.

Raihan, A. 2024. A review of recent advances, challenges, and potential future directions of climate-smart agriculture. In: *Proceedings of the International Conference on Climate-Smart Agriculture*, Bangkok, Thailand.

Reddy, K.S., Pathak, H., Prabhakar, M., Rao, K.V., Rao, C.R. 2018. Innovations for Climate-Resilient Agriculture. *Soil and Water Management Innovations towards Doubling the Farmers' Income.*

Sati, V.P., Vangchhia, L. 2017. *A Sustainable Livelihood Approach to Poverty Reduction: An Empirical Analysis from Mizoram, India.* Springer, New York.

Setboonsarng, S. 2006. *Organic Agriculture, Poverty Reduction, and the Millennium Development Goals* (ADBI Discussion Paper No. 54). Asian Development Bank Institute.

Sharma, V., Tripathi, A.K., Mittal, H. 2022. Technological revolutions in smart farming: Current trends, challenges and future directions. *Comput. Electron. Agric.* 201: 107217.

Shyam, C.S., Shekhawat, K., Rathore, S.S., Subhash, B., Singh, R.K., Upadhyay, P.K., Dass, A., Fatima, A., Kumar, S., Sanketh, G.D., Singh, V.K. 2023. Development of integrated farming system model: A step towards achieving biodiverse, resilient and productive green economy in agriculture for small holdings in India. *Agron.* 13: 955.

Singh, S., Kumawat, K.C. 2023. Integrated farming system for improving sustainability in agriculture. *Ind. Farm.* 73(9): 40–43.

Singh, V.K., Rathore, S.S., Singh, R.K., Upadhyay, P.K., Shekhawat, K. 2020. Integrated farming system approach for enhanced farm productivity, climate resilience and doubling farmers' income. *Ind. J. Agricul. Sci.* 90(8): 1378–1388.

Soni, R.P., Katoch, M., Ladohia, R. 2014. Integrated farming systems-a review. *IOSR-JAVS.* 7(10): 36–42.

Tamil Nadu Agricultural University (TNAU). 2018. *Apiculture.* TNAU Agritech Portal. http://www.agritech.tnau.ac.in/.

The Hindu. 2024. *Tamil Nadu Agriculture Budget Highlights.* https://www.thehindu.com/.

8 Relevance of Quality Assurance Process under Environmental Governance

Mandeep Kaur, Tank Sinderpal Kabalsingh, and Archana Saini

8.1 INTRODUCTION

The environmental governance refers to methods and procedures that ensure environmental data, policies, and activities adhere to specified criteria of accuracy, dependability, and integrity. High-quality data and decision-making procedures are essential for effective governance as environmental concerns become more complicated, encompassing everything from decline in biodiversity and shift in climate due to air and water pollution. Quality assurance (QA) provides the foundation upon which sound environmental management is based and ensures transparency, accountability, and continuous improvement in environmental performance. Fundamentally, QA in environmental governance aims to guarantee that management, reporting, and monitoring initiatives are consistent and supported scientifically (Lemos and Agrawal, 2006). Without QA, environmental decisions can be based on inaccurate or inconsistent data, leading to ineffective or even harmful outcomes. QA makes sure that every step of the process, i.e., from gathering and analyzing data to creating and implementing policies, is carried out in accordance with well-defined. QA is not just a technical or administrative function but a strategic framework that supports effective environmental governance. It aligns that environmental initiatives are both legitimate and responsible by balancing scientific rigor with policy relevance. To achieve sustainability goals, integrating strong QA principles at different levels of governance is required, right from local projects to international treaties, as environmental issues become more urgent and intricate (Jagers and Stripple, 2003). The various principles of QA in environmental governance are discussed in Figure 8.1.

8.2 ROLE OF QUALITY ASSURANCE (QA) IN ENVIRONMENTAL MONITORING AND ASSESSMENT

QA plays a fundamental role in environmental monitoring and assessment by ensuring that data collected from environmental systems are accurate, reliable, and reproducible. Environmental data can be deceptive in the absence of strong QA, which

DOI: 10.1201/9781003452805-8

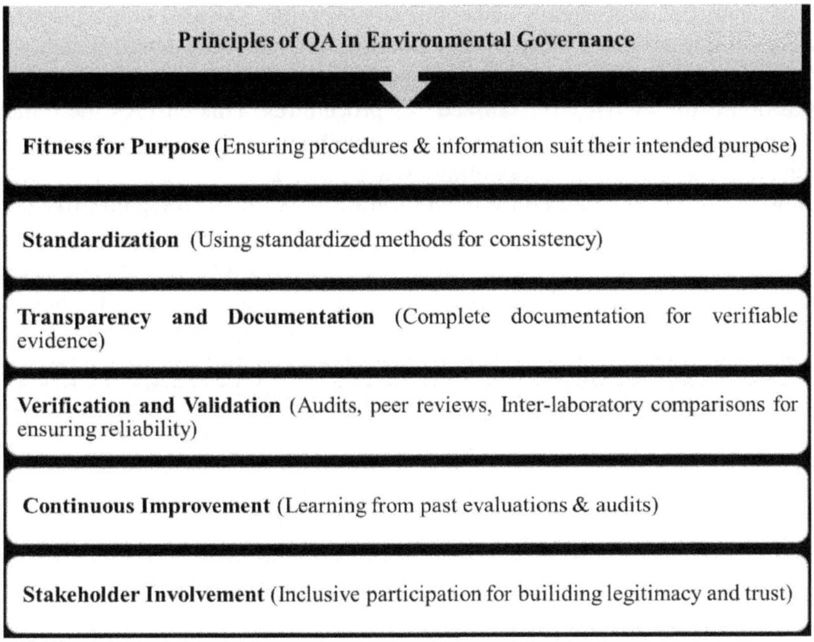

FIGURE 8.1 Principles of QA in environmental governance.

can result in poor decision-making, regulatory failures, and weak public confidence. Environmental monitoring involves the organized gathering of data on various environmental factors, including air and water quality, soil condition, and biodiversity. QA ensures that these data are gathered using validated methods, calibrated equipment, and standardized procedures. According to U.S. Environmental Protection Agency (EPA), QA in environmental monitoring includes activities such as creating quality management plans, conducting performance audits, and validation and verification of data (UNEP, 2022b). These procedures ensures that the data satisfies particular quality standards, such as completeness, correctness, representativeness, and precision.

The most important aspect of QA is the Quality Control (QC) component, which includes operational methods and tasks needed to meet quality criteria. For example, blind sample testing, replicate sampling, and instrument calibration are crucial QC procedures that confirm the integrity of the monitoring procedure (UNEP, 2019). Without these procedures, there is a higher risk of possibility of bias and measurement errors, which can skew environmental assessments.

When monitoring data are used for policymaking and environmental impact assessments, the role of QA becomes even more crucial. For instance, strict QA procedures guarantee that water quality evaluations in all member states are uniform and comparable under the European Union's (EU) Water Framework Directive (WFD). These QA measures support transboundary water management and allow for the benchmarking of environmental performance (European Commission, 2020a). In this situation, QA assists policymakers in establishing suitable environmental goals and assessing the success of implemented measures.

By ensuring data consistency and continuity over time, QA also supports long-term environmental monitoring projects. For example, Global Atmosphere Watch (GAW) program of the World Meteorological Organization's (WMO) monitors atmospheric composition using globally recognized QA procedures. This ensures the reliability of long-term statistics on air pollutants and greenhouse gases, which can be utilized for trend analysis and climate modeling (WMO, 2021).

Additionally, QA is essential for incorporating emerging technologies like drones, automated sensors, and remote sensing into environmental monitoring. In order to ensure the dependability of digital monitoring systems, QA frameworks assist in validating and calibrating the massive amounts of data generated by these technologies. Data interoperability between industries and geographical areas is further facilitated by standardizing QA procedures for emerging technologies. In community-based monitoring, QA protocols help legitimize data collected by nonexperts. Community-sourced data can be improved by guidelines and training, making it appropriate for official evaluations and decision-making (Conrad and Hilchey, 2011).

Undoubtedly, QA is essential to environmental evaluation and monitoring. It encourages public confidence in environmental governance, helps in efficient policymaking, and ensures the scientific validity of environmental data. Investing in strong QA systems will be essential to attain sustainable and knowledgeable environmental management.

8.3 INTERNATIONAL STANDARDS AND GUIDELINES (ISO 14000, EPA, EU DIRECTIVES)

Global, regional, and national environmental governance are shaped and harmonized in large part by international norms and principles. Among the most significant of these are the EU environmental directives, the U.S. EPA rules, and the ISO 14000 series.

Each of these frameworks offers structured methods for pollution management, resource conservation, and sustainable development, and each plays a distinct role in ensuring environmental QA and compliance. When combined, they offer a thorough set of guidelines and procedures that help businesses, governments, and organizations fulfill their environmental obligations. The International Organization for Standardization (ISO) created the widely accepted ISO 14000 series of standards, which are focused on environmental management systems (EMS).

Its most well-known standard, ISO 14001, provides a framework that organizations can follow to set up an effective EMS. The primary goal of this standard is not necessarily to reduce the environmental impact outright but to establish systematic processes that enable continual improvement in environmental performance (ISO, 2023a). ISO 14001 emphasizes the Plan-Do-Check-Act (PDCA) cycle, ensuring that companies plan their environmental objectives, implement them, monitor their progress, and take corrective actions for continuous improvement. By adopting ISO 14001, organizations demonstrate their commitment to comply with environmental legislation, reducing waste, and preventing pollution. Importantly, certification to ISO 14001 has become a market differentiator, enhancing an organization's credibility and reputation among consumers and stakeholders. In addition, ISO 14004 provides broad guidance on the principles, frameworks, and supporting methods for

EMS, while ISO 14031 provides guidance on evaluating environmental performance using relevant indicators. These standards help align corporate operations with global sustainability goals and ensure that environmental management is embedded into the core strategic functions of businesses (ISO, 2023a).

The EPA in the United States serves as the chief federal authority responsible for safeguarding both environment and societal well-being. The EPA creates and implements rules under a range of environmental laws, such as the Comprehensive Environmental Response, Compensation, and Liability Act (CERCLA), the Resource Conservation and Recovery Act (RCRA), the Clean Air Act (CAA), and the Clean Water Act (CWA). EPA's QA standards for environmental monitoring and pollution management are supported by these laws (EPA, 2023a). For example, the CAA empowers the EPA to set National Ambient Air Quality Standards (NAAQS) to limit harmful pollutants such as ozone, sulfur dioxide, and particulate matter, ensuring air quality is maintained at levels protective of public health and ecosystems (EPA, 2023b). The EPA also enforces standards for industrial emissions and vehicle exhaust, utilizing QA protocols like emissions testing, permitting, and performance reviews to ensure compliance. Likewise, the CWA regulates the release of pollutants into U.S. water bodies, mandating that industries secure permits through the National Pollutant Discharge Elimination System (NPDES), which are subject to stringent QA and monitoring requirements (EPA, 2023c). Additionally, the EPA promotes QA through its Quality Program, which provides a structured approach for ensuring that environmental data collected, analyzed, and used in decision-making are scientifically valid and legally defensible. This program includes the development of Quality Management Plans and Quality Assurance Project Plans, which ensure that all environmental data activities meet defined standards of quality and reliability (EPA, 2023d). Through these rigorous regulatory instruments and QA management, air and water quality, and emissions monitoring, and they are central to ensuring uniform environmental standards across the EU's 27 member states.

The Waste Framework Directive (2008/98/EC) encourages extended producer responsibility and the creation of circular economy models while outlining the waste hierarchy: prevention, reuse, recycling, recovery, and disposal (European Commission, 2022a). Additionally, it requires member nations to create waste management programs and plans with quantifiable standards for QA and reporting. The Air Quality Directive (2008/50/EC) sets limit values for major air pollutants such as nitrogen dioxide and particulate matter, requiring regular monitoring, public reporting, and the development of air quality improvement plans in areas exceeding pollution thresholds (European Commission, 2022b).

For water quality, the WFD (2000/60/EC) is one of the most ambitious pieces of EU legislation, aiming for all water bodies, rivers, lakes, coastal waters, and groundwater to achieve "good status" through integrated water resource management and QA-driven monitoring programs (European Commission, 2022c). Additionally, the Industrial Emissions Directive (2010/75/EU) requires that industrial operators obtain permits based on Best Available Techniques, which are reviewed and updated using technical working groups involving stakeholders across the EU. This directive also includes strict monitoring and reporting requirements, ensuring that emissions to air, water, and soil are kept within regulatory limits (European Commission, 2022d).

What distinguishes EU directives from many other regulatory instruments is their strong emphasis on transparency, public participation, and regular revision. For instance, under the Aarhus Convention, which the EU has ratified, public authorities are obligated to provide access to environmental information and involve citizens in environmental decision-making processes. This ensures that QA in environmental governance is not only a top-down administrative function but also a participatory one that engages civil society and promotes accountability (European Environment Agency, 2023). Moreover, the EU utilizes the European Environment Agency (EEA) to collect, analyze, and disseminate environmental data, supporting evidence-based policy development and facilitating cross-border environmental assessments (European Environment Agency, 2023).

Stakeholder involvement in environmental management, scientific integrity, and continual improvement are shared by all three frameworks: the EU directives, EPA rules, and the ISO 14000 series. However, they also reflect differing institutional and legal cultures. While EU directives operate within a supranational governance structure where compliance is guaranteed by legal obligations and possible penalties imposed by the European Court of Justice, ISO standards are voluntary and market-driven, while EPA regulations are statutory and enforceable under U.S. federal law. Despite these distinctions, there is a growing movement toward mutual recognition and interoperability, especially as international development organizations and multinational enterprises look for uniform environmental standards for operations across borders.

8.4 ENVIRONMENTAL MANAGEMENT SYSTEMS (EMS) AND ROLE OF GOVERNMENT REGULATORY BODIES

EMS, particularly ISO 14001, offer a structured framework for organizations to enhance their environmental performance through sustainable practices (ISO, 2023a). At the heart of ISO 14001 is the PDCA cycle, which encourages continuous improvement. During the "Plan" phase, organizations assess environmental impacts, establish legal requirements, and set measurable goals (EPA, 2023g). The "Do" phase involves implementing these goals through resource management, risk mitigation, and training (ISO, 2023a). In the "Check" phase, organizations monitor performance, conduct audits, and evaluate progress (EPA, 2023h). The final "Act" phase focuses on using audit outcomes to guide corrective actions and improve EMS processes (ISO, 2023a). By integrating EMS into daily operations, businesses promote sustainability, operational efficiency, and legal compliance.

ISO 14001 also plays a key role in regulatory compliance, requiring organizations to monitor environmental legislation, maintain legal registers, and perform regular compliance audits (EPA, 2023g). This reduces legal risks and supports responsible business conduct. Through enhanced resource efficiency, such as water conservation, emissions reduction, and recycling initiatives, businesses not only lower their environmental footprint but also achieve cost savings (Smithers, 2024; Barbhuiya et al., 2024). ISO 14001 certification enhances stakeholder confidence and frequently becomes a prerequisite in international supply chains (ISO, 2023a). Employee engagement also improves, as the framework fosters environmental awareness and accountability at all levels (Bishop and Govindarajulu, 2009).

Environmental audits are a critical QA component of EMS, ensuring the system's efficacy and compliance. Internal audits identify inefficiencies and prepare organizations for external evaluations, while third-party audits confirm ISO 14001 adherence and often lead to certification, which enhances corporate credibility (Bansal and Roth, 2000; Carter and Rogers, 2008). These audits follow rigorous QA standards to maintain objectivity and precision (Zutshi and Sohal, 2004).

Despite these advantages, EMS implementation can be challenging, especially for small and medium-sized enterprises, due to significant initial investments and the need for strong leadership commitment. However, consistent training, monitoring, and cultural support can help overcome these challenges and yield long-term environmental and financial benefits (Zutshi and Sohal, 2004). In this context, government organizations and regulatory bodies play a pivotal role in shaping, implementing, and enforcing environmental laws and policies at both national and international levels. These institutions such as the United Nations Environment Programme (UNEP), the U.S. EPA, and national environmental ministries use QA systems to uphold legal compliance, promote sustainability, and improve environmental outcomes (EPA, 2023e; UNEP, 2022a).

Key QA tools used by these bodies include standard-setting, inspections, audits, monitoring systems, and performance reviews. For example, the EPA enforces regulations such as CAA and CWA through emissions permits, regular inspections, and stringent enforcement measures. Noncompliance can result in penalties or operational shutdowns, serving as a deterrent to environmental violations. Internationally, UNEP supports global environmental governance by facilitating policy frameworks, coordinating environmental programs, and ensuring compliance with agreements such as the Convention on Biological Diversity and the Paris Agreement. It leads scientific assessments like the Global Environment Outlook (GEO), which are grounded in QA standards to ensure transparency and reliability (UNEP, 2022a).

National environmental ministries further align domestic laws with global commitments and foster capacity-building through local monitoring and enforcement systems. Countries like Germany and Japan have adopted integrated QA frameworks that blend legal enforcement with voluntary EMS practices, encouraging industries to proactively reduce their ecological footprint (OECD, 2021). QA processes also bolster transparency and accountability, which are essential for public trust. Environmental data publication, stakeholder engagement, and accessible information platforms enable civil society to monitor and influence environmental performance. These practices reinforce democratic governance and stimulate more sustainable actions across sectors (Bäckstrand, 2006).

8.5 QUALITY ASSURANCE IN CORPORATE SOCIAL RESPONSIBILITY (CSR) AND ETHICAL COMPLIANCE

QA also underpins corporate social responsibility (CSR) and ethical compliance by transforming broad commitments into measurable environmental actions. Green supply chain management, a vital aspect of CSR, encourages firms to evaluate suppliers based on environmental standards, emissions, and regulatory compliance. QA systems support this by helping companies establish performance indicators, conduct audits, and foster continual improvement (Welford, 2005). Through employee training, EMS development, and internal audits, QA enables organizations to implement sustainable practices that align with CSR goals such as pollution prevention and resource conservation. QA frameworks help companies reduce indirect environmental impacts and support carbon footprint reduction efforts via renewable energy, process optimization, and waste minimization (Zhu and Geng, 2013).

Third-party audits and global reporting frameworks like the Global Reporting Initiative (GRI) increase CSR transparency. External QA validates sustainability reports, ensures credibility, and identifies improvement areas (GRI, 2021). This integration of QA strengthens environmental accountability, stakeholder trust, and corporate competitiveness.

8.6 PUBLIC PARTICIPATION AND TRANSPARENCY IN ENVIRONMENTAL GOVERNANCE

Public participation and transparency are critical for effective and democratic environmental governance. These principles enhance policy legitimacy, foster public trust, and ensure accountability. As environmental challenges grow in complexity such as climate change and biodiversity loss, there is an increasing need for inclusive, data-driven governance systems.

Transparency begins with open access to environmental data. When governments provide data on air and water quality, emissions, impact assessments, and pollution, they empower citizens, civil society, and researchers to hold decision-makers accountable (Gupta, 2008). This reduces corruption and supports informed public oversight (UNEP, 2021b).

Digital platforms and geospatial technologies have revolutionized transparency. Initiatives like UNEP's Global Environment Monitoring System (GEMS) and the U.S. EPA's Envirofacts database provide public access to comprehensive environmental information (EPA, 2023f; UNEP, 2022b). These platforms enable greater public engagement in decision-making and monitoring.

Meaningful public participation involves genuine engagement in environmental policymaking through hearings, consultations, and community monitoring. Arnstein's (1969) "Ladder of Citizen Participation" stresses the importance of power-sharing between authorities and citizens. Policies shaped by inclusive participation are more equitable, well-informed, and sustainable.

Engaging diverse stakeholders including indigenous communities, industry, academia, and NGOs ensures broader perspectives and more balanced policies. For example, indigenous knowledge contributes to sustainable resource management, while industry stakeholders can offer insights into regulatory feasibility (Reed, 2008). At the international level, frameworks such as the Aarhus Convention mandate access to information, public participation, and justice in environmental matters. These principles are also integral to the Sustainable Development Goals (SDGs), particularly SDG 13 (climate action) and SDG 16 (inclusive institutions) (UNECE, 1998; UNEP, 2021b). Robust EMS, effective regulatory bodies, strong QA systems, corporate responsibility, and active public participation together form the foundation of effective environmental governance. These interconnected elements enhance transparency, ensure compliance, and drive sustainable development.

8.7 HARMONIZING GLOBAL AND LOCAL QA STANDARDS

Aligning international and domestic QA standards is necessary to establish a unified and efficient system of environmental governance. Environmental problems like climate change, pollution, and loss of biodiversity are transboundary in character, and thus uniform standards and concerted regulatory systems are becoming more and more important (Najam et al., 2006). Yet, significant differences between national regulatory systems underpinned by alternative priorities, economic endowments, institutional arrangements, and environmental conditions have tended to result in piecemeal responses. These variations can undermine implementation and compliance, making it easier for transnational businesses to take advantage of loopholes or "shop around" for more relaxed standards. Such regulatory fragmentation weakens the effectiveness of global environmental accords, including the Paris Agreement on climate change and the Basel Convention on hazardous waste control, as well as undermining efforts to enhance environmental accountability and global sustainability.

The absence of harmonized QA standards also makes monitoring, reporting, and cross-country comparison of environmental performance even more complex. It undermines the confidence of priority stakeholders like governments, companies, and civil society entities, and lessens the collective capacity to tackle global environmental issues. Thereafter, a number of international initiatives have been set in motion aimed at harmonizing national QA systems with global QA frameworks. One key contribution from the ISO has been through the creation of the ISO 14001 standard, which provides an internationally accepted framework for EMS (ISO, 2023b). ISO 14001 gives directions for establishing environmental goals, internal auditing, and continuous improvement, while its flexible structure accommodates flexibility for application in diverse local conditions without compromising on fundamentals.

In addition, capacity-building initiatives by the UNEP are essential in supporting developing nations in developing their environmental QA systems. UNEP activities—technical assistance, training, institutional strengthening financing, and national environmental information system development—are aimed at improving the capacity of local governments to effectively implement and enforce QA systems and tailor them to their respective socioeconomic and ecological realities (UNEP, 2021a).

Harmonization of QA standards is also advantageous for multinational companies. Uniform environmental requirements across nations reduce the cost of compliance, clarify reporting requirements, and provide equivalence in environmental performance along global value chains. Improved consistency enhances transparency and comparability of environmental information, leading to greater accountability before regulators, investors, and wider society.

8.8 CASE STUDIES OF QUALITY ASSURANCE SUCCESS IN ENVIRONMENTAL GOVERNANCE

QA is fundamental in environmental governance to ensure that policies and actions meet established standards and achieve intended outcomes. Successful case studies show that effective QA measures can enhance environmental monitoring, regulatory compliance, and sustainable development. Mechanisms based on community QA have also become prominent. Participatory monitoring programs in forest conservation enable local communities to monitor deforestation, evaluate biodiversity, and test the effectiveness of REDD+ programs (FAO, 2016). Hence, even though environmental tragedies highlight the devastating effects of QA failures, they have also pushed through major innovations. Adopting technological innovations and participatory means can significantly boost QA systems and fortify actions in climate change mitigation and sustainable environmental management. The key examples in this context are discussed in Table 8.1.

TABLE 8.1
Global Case Studies Illustrating the Successes, Failures, and Innovations in Quality Assurance (QA) Practices within Environmental Governance

Category	Case Study/ Program	QA Measures/ Failures	Impact/Outcome	References
✓ Success	EU Water Framework Directive (WFD)	Standardized reporting and monitoring, harmonized QA practices	Improved water quality, reduced pollution, enhanced cross-border cooperation	European Commission (2020b)
✓ Success	U.S. EPA—Air Quality System (AQS)	Instrument calibration, personnel training, performance testing	Reduced SO_2 and NOx levels, enhanced air quality data reliability	U.S. EPA (2022a)
✓ Success	India—National Air Quality Monitoring Programme (NAMP)	Equipment calibration, interlaboratory checks, Standard Operating Procedure (SOP) adherence	Improved data integrity, supported National Clean Air Programme (NCAP) goals to reduce PM2.5/PM10	CPCB (2019)
✓ Success	Australia—National Pollutant Inventory (NPI)	Automated and manual QA checks, public validation	Increased transparency, informed policy and industrial compliance	Australian Government (2021)

(Continued)

TABLE 8.1 (*Continued*)
Global Case Studies Illustrating the Successes, Failures, and Innovations in Quality Assurance (QA) Practices within Environmental Governance

Category	Case Study/ Program	QA Measures/ Failures	Impact/Outcome	References
✓ Success	IPCC Climate Assessments	Peer review, data validation, transparency provisions	High credibility reports, supported global agreements like Paris Agreement	IPCC (2021)
✓ Success	Canada—National Pollutant Release Inventory (NPRI)	Mandatory QA protocols for industry reporting, periodic audits, and public disclosure tools	Improved pollution transparency and public accountability	Environment and Climate Change Canada (2020)
✓ Success	Germany—Blue Angel Eco-label	Independent certification, criteria updates, lifecycle analysis	Promoted environmentally friendly products and raised consumer awareness	Federal Ministry for the Environment (2018)
✗ Failure	Bhopal Gas Tragedy (India, 1984)	Poor maintenance, absent safety protocols, failed risk assessment	Thousands of deaths, long-term health/ environmental damage	Dhara and Dhara (2002)
✗ Failure	Deepwater Horizon Oil Spill (USA, 2010)	Poor training, equipment failures, weak safety oversight	Massive ecological damage, legal and regulatory reforms	National Commission (2011)
✗ Failure	Fukushima Daiichi Nuclear Disaster (Japan, 2011)	Inadequate disaster planning, weak QA protocols	Reactor meltdowns, international nuclear safety reforms	IAEA (2015)
✗ Failure	Minamata Mercury Disaster (Japan, 1950s)	Neglect of industrial discharge controls, lack of early QA monitoring	Widespread mercury poisoning, led to international conventions on mercury	UNEP (2013)
↻ Innovation	Measurement Reporting and Verification (MRV) under Paris Agreement	Standardized reporting, remote sensing, AI-driven QA	Improved emission tracking and transparency in climate commitments	UNFCCC (2020)

(*Continued*)

TABLE 8.1 (*Continued*)
Global Case Studies Illustrating the Successes, Failures, and Innovations in Quality Assurance (QA) Practices within Environmental Governance

Category	Case Study/ Program	QA Measures/ Failures	Impact/Outcome	References
Innovation	Blockchain in Carbon Markets	Tamper-proof records, real-time tracking	Reduced fraud in carbon offsets, increased trust and transparency	Dutta and Roy (2021)
Innovation	Community QA in REDD+ Programs	Participatory monitoring, local biodiversity assessments	Empowered communities, improved forest conservation efforts	FAO (2016)
Innovation	Earth Observation for Sustainable Development (EO4SD)	Satellite-based QA monitoring, standardized remote sensing analytics	Enhanced environmental monitoring in developing countries	European Space Agency (2021)
Innovation	Smart Sensor Networks in Urban Air Monitoring (e.g., Barcelona)	Real-time data QA, AI-driven data validation	Increased spatial resolution of air quality data and public access	Smart Citizen Project (2022)

8.9 CONCLUSION

Environmental governance is a complicated process that involves many different institutions, multiple factors, and overlapping legal systems. As the environmental crisis deepens due to biodiversity loss, pollution, climate change, and unsustainable resource exploitation, the need for robust, transparent, and effective governance mechanisms becomes more important than ever. This chapter confirms that incorporating QA into environmental governance is a fundamental element of a robust governance architecture rather than an add-on feature. The conclusion drawn from this chapter highlights that QA serves as the backbone of trust, accountability, and efficiency in environmental governance.

Governments and organizations may guarantee that environmental policies are not only well-intentioned but also effective in their implementation by instituting systematic checks and balances, including thorough environmental impact studies, public consultations, and scientific audits. QA mechanisms encourage stakeholder confidence, promote policy coherence, and enable the discovery and rectification of governance shortcomings in a timely way. With strong environmental regulations incorporated in the Constitution and progressive judicial interpretations, India boasts a comprehensive legal framework. However, political opposition, bureaucratic delay, and enforcement gaps frequently hinder the effectiveness of these legislation. Thus, QA becomes a vital instrument for bridging the gap between practice and policy.

The chapter also highlights how crucial community involvement and decentralization are to environmental decision-making. When local communities, indigenous populations, and civil society organizations are empowered to make meaningful contributions, inclusive governance leads to true QA. Through information sharing, public hearings, and participatory monitoring, environmental governance becomes more democratic. As seen by ISO 14001 certifications, ESG indicators, and transboundary protocols, there is a growing global trend toward standardizing QA. Adopting global best practices gives nations like India a road map for creating responsive and resilient institutions. In essence, QA is both a practice and a philosophy—anchoring trust, accountability, and proactive sustainability for future generations.

REFERENCES

Australian Government. (2021). *National Pollutant Inventory*. https://www.npi.gov.au.

Bäckstrand K. (2006). Multi-stakeholder partnerships for sustainable development: Rethinking legitimacy, accountability and effectiveness. *Environ Policy Gover.*, 16(5), 290–306.

Bansal P. and Roth K. (2000). Why companies go green: A model of ecological responsiveness. *Acad. Manag. J.*, 43(4), 717–736.

Barbhuiya S., Das B.B. and Adak D. (2024). A comprehensive review on integrating sustainable practices and circular economy principles in concrete industry. *J. Environ. Manage.*, 370, 122702.

Bishop J.W. and Govindarajulu N. (2009). A conceptual model for organizational commitment to environmental practices. *Bus. Strategy Environ.*, 18(8), 453–465.

Carter C.R. and Rogers D.S. (2008). A framework of sustainable supply chain management: Moving toward new theory. *Int. J Phys. Distrib. Logist. Manag.*, 38(5), 360–387.

Central Pollution Control Board (CPCB) (2019). National Air Quality Monitoring Programme. https://www.cpcb.nic.in.

Conrad C.C. and Hilchey K.G. (2011). A review of citizen science and community-based environmental monitoring. *Environ. Monit. Assess.*, 176: 273–291.

Dhara V.R. and Dhara R. (2002). The union carbide disaster in Bhopal: A review of health effects. *Arch. Environ. Health*, 57(5): 391–404.

Dutta S. and Roy S. (2021). A review of blockchain technology for environmental sustainability. *Sustain. Dev.*, 29(3): 548–563.

Environment and Climate Change Canada. (2020). National Pollutant Release Inventory (NPRI) Annual Report. https://www.canada.ca/en/services/environment/pollution-waste-management/national-pollutant-release-inventory.html.

Environmental Protection Agency (EPA). (2023a). Environmental Laws and Regulations. https://www.epa.gov/laws-regulations.

Environmental Protection Agency. (EPA) (2023b). Clean Air Act Standards and Guidelines. https://www.epa.gov/clean-air-act-overview.

Environmental Protection Agency. (EPA) (2023c). National Pollutant Discharge Elimination System (NPDES). https://www.epa.gov/npdes.

Environmental Protection Agency. (EPA) (2023d). Quality Program for Environmental Data Collection and Use. https://www.epa.gov/quality.

Environmental Protection Agency. (EPA) (2023e). About the U.S. Environmental Protection Agency. https://www.epa.gov/aboutepa.

Environmental Protection Agency (EPA) (2023f). Envirofacts Database. https://www.epa.gov/enviro.

Environmental Protection Agency (EPA) (2023g). Environmental Management Systems. https://www.epa.gov/ems.

Environmental Protection Agency. (EPA) (2023h). EMS Audit Protocol. https://www.epa.gov/compliance/audit-protocols.

European Commission. (2020a). Water Framework Directive and QA in Water Monitoring. https://ec.europa.eu.

European Commission. (2020b). Water Framework Directive and the Floods Directive. https://ec.europa.eu.

European Commission. (2022a). Waste Framework Directive. https://environment.ec.europa.eu/topics/waste-and-recycling/waste-framework-directive_en.

European Commission. (2022b). Air Quality Standards. https://environment.ec.europa.eu/topics/air/air-quality/eu-air-quality-standards_en.

European Commission (2022c). Water Framework Directive. https://environment.ec.europa.eu/topics/water/water-framework-directive_en.

European Commission. (2022d). Revised Industrial Emissions Directive and Regulation Establishing the Industrial Emissions Portal: outcomes and opportunities. https://eeb.org/wp-content/uploads/2024/08/IED-and-IEP-R-assessement.pdf.

European Environment Agency (EEA). (2023). Environmental Data Centre. https://www.eea.europa.eu/en/datahub.

European Space Agency. (2021). Earth Observation for Sustainable Development (EO4SD). https://eo4sd.esa.int/.

FAO. (2016). Community-based Monitoring and QA in Forest Governance. https://www.fao.org.

Federal Ministry for the Environment. (2018). Blue Angel—the German Ecolabel. https://www.blauer-engel.de/en.

Global Reporting Initiative (GRI). (2021). Global Reporting Initiative Standards. https://www.globalreporting.org.

Gupta A. (2008). Transparency under scrutiny: Information disclosure in global environmental governance. *Glob. Environ. Polit.*, 8(2): 1–7.

Intergovernmental Panel on Climate Change (IPCC). (2021). AR6 Climate Change 2021: The Physical Science Basis. https://www.ipcc.ch.

International Atomic Energy Agency (IAEA). (2015). *The Fukushima Daiichi Accident*. IAEA, Vienna.

ISO. (2023a). ISO 14001: Environmental Management Systems. https://www.iso.org.

ISO. (2023b). ISO 14001: Environmental Management Systems—Requirements with Guidance for Use. https://www.iso.org/iso-14001-environmental-management.html.

Jagers S.C. and Stripple J. (2003). Climate governance beyond the state. *Glob. Governance*, 9, 385–399.

Lemos M.C. and Agrawal A. (2006). Environmental governance. *Annu. Rev. Environ. Resour.*, 31, 297–325.

Najam A., Huq S. and Sokona Y. (2006). Climate negotiations beyond Kyoto: Developing countries' concerns and interests. *Clim. Policy*, 5(2): 153–175.

National Commission on the BP Deepwater Horizon Oil Spill. (2011). *Deep Water: The Gulf Oil Disaster and the Future of Offshore Drilling*. Government Printing Office, Washington, DC.

Organisation for Economic Co-operation and Development (OECD). (2021). *Environmental Performance Reviews: Towards Better Environmental Quality*. OECD Publishing, Paris, France.

Reed M.S. (2008). Stakeholder participation for environmental management: A literature review. *Biol. Conserv.*, 141(10): 2417–2431.

Smart Citizen Project. (2022). Smart Air Quality Monitoring in Urban Spaces. https://smartcitizen.me/.

Smither. (2024). The Benefits of ISO 14001 for Your Organization. https://www.smithers.com/resources/2024/july/the-benefits-of-iso-14001-for-your-organization.
UNEP. (2019). Guidelines for Environmental Monitoring and Reporting. UNEP, Nairobi, Kenya.
UNEP. (2021a). Environmental Governance: Capacity Building Initiatives. https://www.unep.org.
UNEP. (2021b). Making Peace with Nature: A Scientific Blueprint to Tackle the Climate, Biodiversity and Pollution Emergencies. UNEP, Nairobi, Kenya.
UNEP. (2022a). United Nations Environment Programme Annual Report. https://www.unep.org/resources/annual-report-2022.
UNEP (2022b). GEMS/Water Programme. https://www.unep.org/resources/gems.
United Nations Economic Commission for Europe (UNECE). (1998). Convention on Access to Information, Public Participation in Decision-Making and Access to Justice in Environmental Matters (Aarhus Convention). UNECE, Geneva, Switzerland.
United Nations Environment Programme (UNEP). (2013). Global Mercury Assessment 2013: Sources, Emissions, Releases and Environmental Transport. UNEP, Nairobi, Kenya.
United Nations Framework Convention on Climate Change (UNFCCC). (2020). MRV Framework under the Paris Agreement. https://unfccc.int.
Welford R. (2005). Corporate social responsibility in Europe, North America and Asia. *J. Corp. Citizenship*, 17: 33–52.
World Meteorological Organization (WMO). (2021). Global Atmosphere Watch (GAW) Programme. https://public.wmo.int.
Zhu Q. and Geng Y. (2013). Drivers and barriers of extended supply chain practices for energy saving and emission reduction among Chinese manufacturers. *J. Clean. Prod.*, 40: 6–12.
Zutshi A. and Sohal A.S. (2004). Adoption and maintenance of environmental management systems. *Manag. Environ. Qual. An Int. J.*, 15(4), 399–419.

9 Sustainable Animal Feed for Safer Animal Farming

Amro Abd Al Fattah Amara,
Mounir Mohamed Salem-Bekhit,
Khairy Mohamed Abdallah Zoheir,
Nawal Abd El-Baky,
Mohamed Morsi Mohamed Ahmed,
and Arti Gupta

9.1 INTRODUCTION

The quality of the animal food is an important factor in any animal farming process, as it can be a source of different kinds of infections, chemical residues, toxins, etc. Ensuring that the quality of animal foods is controlled is insufficient to prevent the spread of microbes, parasites, ticks, cysts (worms), etc. Every action could, intentionally or non-intentionally, affect the food quality, which in turn may have an impact on the welfare and health of the animals (Broom, Galindo, and Murgueitio 2013; Phillips 2024), the products' quality, the consumers' health, and other factors. Finally, the effects will be transferred to our food (Figure 9.1). This chapter is concerned with discussing some sustainable practices to improve the quality of animal farming. They are mainly examples of some common real-negative practices, the social and economic roles, the animals' foodborne diseases, the effect of the use of antibiotics and other antimicrobial agents, the animals' hygiene (Tuma 2024), and some animal–human interactions. Some of those items are very close to each other. This chapter attempts to broaden the perspective that has been addressed for several years about the animal's welfare, the food quality, and the reasons that food and other welfare issues are so interconnected and that issues should be balanced. Any deviation from the biological or life balance for any creature will cause negative consequences that range from health degradations to fatal disease(s) (Amara 2013). Animal welfare is a key component of any economy (Meijer et al. 2023; Amara and Gupta 2025), society, environment (Alfonso et al. 2023; Li et al. 2017; Ruckli et al. 2024), farming animals' sustainability (Broom 2010; Crovato et al. 2022), and other factors. Respectable animal welfare consists of healthy, safe food (Leroy et al. 2022); safe, secure, clean, uncrowded housing (Aguayo-Ulloa et al. 2014); proper day/night-light system (Erensoy et al. 2021); no need for drugs or prophylactics; no use of hormones or growth promoters (Colpoys et al. 2019; Sabioni et al. 2022; Satoh et al. 2024; Rashid et al. 2020); appropriate living conditions; natural farming practices and interactions with other animals and the environment (Nair et al. 2021);

 DOI: 10.1201/9781003452805-9

FIGURE 9.1 Timeline of the animal domesticating and farming through time, the development of animal farming, the use of antibiotics, antibiotic resistance, the important points that influence the animal farming style, the First and the Second World Wars, WHO, FAO, and the accumulated problems due to the new crowded fast farming style. (Generated in this study.)

proper social and welfare conditions (Federico and Banks 2020); and protection from any source of illness (Diaz Vicuna et al. 2024) to prevent the need for unwanted drug treatment. In case of infections or any kind of illness, animals need a system for early detection and appropriate treatment. In fact, animals need more care than humans. This is because while humans can save themselves from many effective negative conditions, take decisions (Baldin et al. 2021), and think correctly, animals cannot.

Organic is a term used to describe something being or relating to or derived from or having properties characteristic of living organisms. *Organic Livestock* production is a new term similar in its concept to the term used with the food *Organic food* (Nalubwama, Mugisha, and Vaarst 2011; van Wagenberg et al. 2017), which is a means of food production from farming animals with a large number of rules directed toward a high status of animal welfare, care for the environment, restricted use of medical drugs, and the production of a healthy product without residues (insecticides or chemical residues) (Kijlstra and Eijck 2006). The environmental impact of food of animal origin is currently quantified by the so-called CO_2eq-footprints (Carbon Footprints: CF) (Flachowsky 2011). Toward the sustainable welfare of animal-source food, the concept of animal farming with special care could include other important terms such as *One Health* (health and safety of animals and humans) (Fiorilla et al. 2024; Huang et al. 2020; Mantovani et al. 2022; Shurson 2025; Shurson and Urriola 2022; Shurson, Urriola, and Hung 2022; Shurson, Urriola, and van de Ligt 2022; Thakur et al. 2025; Wang, Hou et al. 2024; Scholten et al. 2013); *Customized Care* (Scholten et al. 2013) (from the perspective of the integrity of the animal); and *No Nuisance* (with regard to environmental and societal perspectives) (Scholten et al. 2013).

Good animal welfare contributes to improving animal food (Nikmaram and Rosentrater 2019; Qiao et al. 2022; Samat et al. 2020), health (El-Sabrout, Khalifah, and Mishra 2023; Luise et al. 2022; Ogbuewu and Mbajiorgu 2023; Saengrung et al. 2023; Samat et al. 2020), and productivity (Kelly et al. 2017). It is also an ethical responsibility (Decuypere et al. 2010; Delvendahl, Rumpold, and Langen 2022; Leroy et al. 2022; Siddiqui et al. 2024). A *One Health* approach (Mantovani et al. 2022) is required, but could be accomplished only when experts from different fields are gathered to take the right decision, implement the right plan, think about the future of animal farming, including which practices can be taken from the past and which should not be used today, and what the future might hold. The multidisciplinary approaches (Randazzo et al. 2020; Zarantoniello et al. 2020) and assessment techniques (Laso et al. 2018; Varela-Ortega et al. 2022) could give a better understanding of the welfare (Siegford 2013).

Raising awareness for the farming animals show correct impacts on animal and human health, as well as of the various forms of risk across levels and institutions, on the environment and economy. Consumers have become more conscious about the negative effects of animal-sourced food consumption on their health and the environment. Consumers who are aware of what is happening in the farms, companies, and markets will be able to identify any unfavorable aspects of the overall process. If the producer succeeds in choosing the best animal products, it may improve the welfare of the animals by encouraging their producers to maintain their products' quality (Vyas et al. 2020). In contrast, avoiding low-quality purchases will encourage producers to enhance the quality of their products. Investment in farming animals (Herrero et al. 2010; Schrobback et al. 2023) can benefit from this awareness by linking producers who adopt sustainable practices (Greenwood 2021) to demand for sustainable products. By increasing consumer awareness, producers will be encouraged to demand support from the government for improving the welfare of the animals and the processes that require it, such as veterinary care, vaccinations, exporting specialized equipment, distributing documentation (Luo et al. 2011; Opoola et al. 2022) to increase governmental and public awareness (Broom 2010; Cornish, Raubenheimer, and McGreevy 2016; Naab et al. 2021), laws, regulations (Żuk-Gołaszewska et al. 2022), and instructions. Martelli (2006) highlights that consumer has a good perception about the link between food quality, food safety, and animal welfare (*from farm to fork*) (fork means a utensil with two or more prongs, used for serving or eating food) (Martelli 2009; Stevenson et al. 2007).

Historically with some exceptions, humans did not experience such a high population in certain narrow areas like the megacities. Only after/during the Industrial Revolution and the First and Second World Wars, millions of workers, soldiers, travelers, and city inhabitants were in demand for an extra amount of food to be available. Even those who are not in special cases (war, travelers, etc.) eat foods that were not prepared at home (Figure 9.1). As a result, the industrialized cities became a place for food factories, restaurants, food markets, food subsidiary products, etc. Industries are in need of producing a product! In other words, if a company produces a certain amount/year, it will increase the total amount of animal food consumed, which could disrupt both the family farming style and the human food supply. If the produce quantities are not sealed in the local market, they will probably find their way to the exo-market (other countries). That means such companies will be a source for evacuating the products that are required for local animals or for human consumption! Sealing products in international markets will be mostly more profitable, and resource consumption will increase. The national planner must not be disconnected from those who are considered as international planners. Succeeding in one sector is not enough.

Other catastrophic attitudes did not come from the governments and their planners, but came from the normal individuals. For example, some experts expected that the number of cultivated areas worldwide should be increased to satisfy such a huge demand for food, from individuals, animals, *exo* demand, etc. But controversially, farmers tend to modernize their lifestyle and use luxury, cultivated areas and grazing areas are abandoned, cities expand into the countryside (or both expand until they meet each other in certain points), and new roads are allocated in the countryside, which engulf more and more of the cultivated areas. The cultivated lands decrease dramatically and the problem becomes more and more complicated. In one day, because of an unexpected war, many countries discovered that they are not able to produce enough food from their land products, but must import food from other countries. Electronics, computers, cars, etc., could not feed humans, particularly in certain cases. Food prices were augmented, and many to survive have gotten rid of some of their animals or at least some of them.

Consumer demand may also inspire political support concerning some farming issues (Brice et al. 2022). It is a sign of quality when producers, the general public, customers, etc. approve the use of restrictive standards and costly processes; ideally, this is done without discriminating against or classifying any human or animal. But roles must be applied wisely to avoid any kind of adverse impact on the producer incomes. In case of small and medium-sized producers, locating international recognition standards can facilitate trade (Al-Mutairi et al. 2023; Ashley et al. 2018; Fatumah et al. 2023; Goumeida et al. 2024; Tizard et al. 2016). Strong political contracts around the importance and urgency of sustainable farming animals' welfare practices are vital for balanced assessment of information and governmental support to local pilot programs and research. When someone initiates the process properly, others will follow suit.

Some of these consequences will be highlighted in this chapter. It concerns the sustainable animal feed for safer animal farming, which is a key point where if one succeeds in applying it correctly, it will guarantee at least healthy humans and animals, animals' welfare, and safe farming conditions for all and for nature. The other social, economic, and political factors will not be discussed in depth but will be highlighted with the exception to those points related directly to the sustainable animal feed for safer animal farming, for example, the welfare of animals and animal producers, the health of consumers, the sustainability of food production, the global environment, and world hunger and food security (Fraser 2001).

The interplay between animal health and food safety can be understood from the perspective of some key points addressed in this chapter. They include the following: (1) The common real-negative practices that could be applied to the animals during different farming styles; (2) the social and economic roles that affect the animal food's amount and quality; (3) on-farm food; (4) the effect of the use of antibiotics and other antimicrobial agents (Amara et al. 2024; Amara and Shibl 2015; Hussein and Amara 2006); (5) animals' hygiene and its related issues (Tuma 2024); and (6) animal–human interactions.

9.2 THE COMMON REAL-NEGATIVE PRACTICES THAT COULD BE APPLIED TO THE ANIMALS DURING DIFFERENT FARMING STYLES

Historically, there have always been kind farmers who take care of their animals, their lives, and their health. But there are many who treat them unwisely and aggressively. For example, some farmers force animals to eat or drink more and more. One old technique is by mixing salt with the animal's food (non-accepted practices by the society) to force the animals to drink more water, aiming to gain more weight. Some do that the day before sealing the animal to gain some extra weight. Some, when sealing a cow or buffalo for milk production, let the milk of the night and the morning to accumulate in the animal's udder; so, it appears with a big udder full of milk. Some use hormones to give fast weight gain, which, instead of increasing the meat content, gives a large amount of fat in the animal's body. The previous practices are usually applied by individuals. But other practices are applied to follow certain policies by big animal farming producers. Some remove the animal's sexual organs to gain more weight. Others castrate the sexual organs (sterilizing) to prevent others from farming certain animals on their own farm (the animals become sterile).

In many cases, animals are used instead of humans to prove or disprove the food quality and safety, and in some cases, they are used as the victims instead of humans. Many accept this concept, because if the animals were not in the front line, humans will be instead. Perhaps that might be acceptable in a narrow range for testing some new drugs. But it must happen under the full control of experts with clear ethical roles. Some animal producers feed animals with foods that might be unacceptable based on their quality or are human food products that have already expired or, in the best scenario, on their way to expire. Agricultural- and food-industry residues comprise a major portion of (approximately over 30%) worldwide agricultural productivity (Ajila et al. 2012). The famous animal that could eat any garbage, waste, expired food, etc., is the pig. Most of these expired foods contain preservatives, toxins, colored chemical compounds, etc. Yes, in the classical family farm, humans used to give animals/birds leftover food from food preparation (e.g., rest of cooked rice to the chickens). But such food is the rest of our fresh, healthy food. In some cases, even the food might contain low levels of undesirable compounds or elements, but such food can still harm the animals if consumed for a long time. Feeding foods that are incompatible and undesired for the animal's digestive system (e.g., ruminant animals) can cause serious illness such as mad cow disease. The outbreak increased throughout the UK because of the practice of feeding meat-and-bone meal to calves of dairy cows (Nathanson, Wilesmith, and Griot 1997).

Improper poultry farming conditions include the day/night (light) (Erensoy et al. 2021), cages (poultry, fish) (Alfonso et al. 2023), antibiotics, hormones, feed style and contents, and farming conditions, which affect the animal welfare and the quality of the final products (Hocquette et al. 2005). Therefore, the chickens do not stop eating day or night as long as they see food in front of them (Amara and Gupta 2025). Calves suffer from drinking their mothers' milk, which is important for their health (Karlsson et al. 2020). The solution will be simple, by letting calves feeding enough milk from their own mothers.

Now, in many places, the farming business has the priority over the allocated amount of area for human food. The other face concerns with the profit (Dartt et al. 1999; Erickson et al. 2020; Fu et al. 2023; Hanson et al. 1998; Heinrichs et al. 2020; Marques et al. 2020; Mohd Azmi et al. 2021; Opoola et al. 2022; Rawski et al. 2021; van Houtert and Sykes 1999; Warner et al. 2022) which could affect the animals' welfare and the final product quality (Gasco et al. 2019; Halmemies-Beauchet-Filleau et al. 2018). Poor awareness of animals and product handling at the farm level can lead to health risks, mortality and morbidity (Paiva et al. 2023; Scott et al. 2019; Will et al. 2024), economic losses (Abdisa and Dilbato 2024), and wasted valuable natural resources (Preston 2009) that challenge the entire supply chain. As such, investors in food production, in particular animal-sourced foods, have key responsibilities in ensuring a proper risk analysis is conducted. All the above-mentioned factors and the other unmentioned factors are correlated with more than one point of view. One factor could be of great interest than the other one. For example, Hocquette et al. (2005) highlighted that consumers' expectations are the basis of the quality definition, food quality is more and more a social concept, and eating quality is an important determinant of food choice (Hocquette et al. 2005). It is the role of the experts from different scientific, technical, practical, political, and social domains to navigate till they reach the best solution for a particular case in a particular area. Human experience, knowledge, intelligence, and the like might fasten any process in its new phase to save a huge amount of allocated investment while the product is a living creature (in its base) and the consumer is the living creature as well.

9.3 THE SOCIAL AND ECONOMIC ROLES THAT AFFECT THE ANIMAL FOOD'S AMOUNT AND QUALITY

9.3.1 Control the Animal Weight

Perhaps the most common problem in the modern farming style is the interest of producers to feed the animals more and more to gain extra weight. Many animals face risks related to excess weight. This chapter will not address all the methods that farmers use to force the animals to gain weight, but some will be addressed. In some cases, the weight loss could not be repaired by giving more food. Early factor side effects could not be recognized in a timely manner, for example, during the lactation period (Santos et al. 2018). The main reasons for keeping animals at a healthy and proper weight include (1) prolonging their lifespan and improving their overall welfare; (2) lowering the chance of developing diabetes (Robertson 2023), high blood pressure (Takayasu et al. 2006; Yeh et al. 2020), kidney problems (Sørensen et al. 2020; Stenberg et al. 2019; Zargari et al. 2024), breathing issues (Melchior, Sève, and Le Floc'h 2004; Zong et al. 2022), and certain other types of illness; (3) lowering the risk of injuries associated with excess weight affecting bones (Muir et al. 2022), joints, and muscles (Wang, Wang, et al. 2024); and (4) helping manage discomfort from arthritis as the animal grows older (Muir et al. 2022; Yu, Lee, and Morota 2021; Yang et al. 2024).

9.3.2 Process Improvement

Protecting, preserving, regarding, enhancing, etc., are all human attitudes to improve themselves, their animals, tools, housing area, etc. For example, food is an essential aspect of life for both humans and animals. Due to many issues, like seasonal variation (Cooke et al. 2025; Maleko et al. 2018), quality of the landscape (Makkar 2018), and the water resources evaluability (Pahlow et al. 2015), humans used to store some of their food and animals' food for times when food is unavailable. For example, humans have learned to preserve their foods by applying different simple techniques like smoking (Dias et al. 2020; Lingbeck et al. 2014; Liu et al. 2018), salting (Amara 2023), drying (Amara 2023), fermenting (Amara 2023), etc. They have done that for their food and their animals' food. They learned to protect the seeds in special containers, drying the edible plants that are suitable for animals' feed, keeping enough salt in the house, etc. For example, in Egypt, trefoil is widely dried to be consumed later by the animals.

9.3.3 The Family Farm: Why Are We All Like It? And Like Its Products?

Over hundreds of years, what we call "the family farm" was optimized from many points of view. For more details about the family farm, refer to Amara and Gupta (2025). New competitors to the family farm appear; such competitors intentionally highlight to us the importance of the family farm and its positive points. The new farming styles appear (Amara and Gupta 2025) with some new general features such as crowded and fast farming cycle; special food formula is produced to optimize the animal growth and weight, or optimized to give more products (Hadfield et al. 2021) from a specific target like eggs (Dawood et al. 2024; Wang et al. 2021; Zhang et al. 2024), milk (Liu et al. 2020; Winkelman, Elsasser, and Reynolds 2008), milk products (Dong et al. 2023), meat (Gill 1999), etc. In contrast to the family farm which run smoothly in most cases, the modern farming style that gathered a huge number of animals and a huge output products forces special and specific practices like the use of (1) preservatives (even bio-preservative) (Ebrahimi et al. 2017; Ray 1992; Ray and Sandine 1992; Sturino, Rajendran, and Altermann 2013), antibiotics (Desbois et al. 2021), and antimicrobial agents (Amara et al. 2024; Amara 2023); (2) hormones (Arieli et al. 2001); and (3) proteins (in different forms) (Eugenio et al. 2022; Øverland, Mydland, and Skrede 2019; Seelam et al. 2022), amino acids (Eugenio et al. 2022; Graham-Thiers and Bowen 2024; Poosuwan, Bunchasak, and Kaewtapee 2010; Spinler et al. 2024), and vitamins (Attia et al. 2023; Wertz et al. 2004; Zhao et al. 2016). Step by step, the animals were treated in crowded fast bulk conditions with all their negative consequences (Amara and Gupta 2025).

In the family farm, animals are used to help in the farm duties, to be used as a source of food for their producer (small family), or to be sealed in a narrow range (Amara and Gupta 2025). Those who work in the daily farm duty or produce other products from cows, buffalo, camels, sheep, and goats are gaining more respect and care. Others are used for protection like dogs and cats or for enjoyment like horses. In addition, one should recognize that animals in many parts of the world are essential for daily activities, e.g., transportation such as horses, donkeys, mules, lamas, and camels. In isolated areas, some animals are part of human survival, like the camels in the deserts. As a result, we should remark that we did not have a fixed interest for a certain animal, but each kind of farming style has one or more than one animals that are considered as important. Even, each type could vary from farm to farm and from location to location. The used tools and technology are another issue. Different categories of tools and technologies are used in different farm styles that differentiate them from (1) primitive and isolated farming area(s); (2) semi-primitive farming conditions where some materialization tools are existed; (3) modern farming area; and (4) advanced modern farming areas.

Animals such as buffalo, cows, sheep, horses, donkeys, dogs, and cats and birds exist in nearly every family farm. Animals that are essential in the farming duties were given enough food and were not exhausted by extra work; their farming places are kept dry, clean, well aerated, and healthy. They are grown with more concern, because any issue in their health and power will block some other essential activities that could corrupt the overall farming processes. Such activities include helping in land irrigation and plowing the ground. Some modernization is affected indirectly in the position of the farming animals as an important part of the farm. After using the electric and machine-dependent irrigation, some are getting rid of their buffalos and cows, because machines are used instead of them. This creates a dramatic effect in the family farm's total products such as milk.

In a more specific analysis of the situation, each of the animals, birds, fish, etc., as well as the human himself who becomes a victim of one or more than one of the many side effects of the changes that happened in the food production cycle concerning animal feeding. But at this specific point, we all discovered that we are nearly corrupting the most solid part that survives humans over the ages, which is the family farm. Over the ages, the family farms have adapted to survive against different problems. In contrast, if companies do not succeed in getting a respective profit, they will close their business (local or international). Modern electric and machine-dependent farmers will stop if no electricity or petroleum exists (for any purpose). What could only survive? Again, it is the traditional farming style where animals do the work, produce the products, and consume only a known amount of food that has never affected the amount of other foods produced for humans. This case could be summarized using the Arabic proverb, "keep running, ships run." Did we keep the running family farm run? And why not?

9.4 ON-FARM FOOD

On-farm food safety and inspection are critical to prevent foodborne diseases, build resilient agri-food operations, and protect producers and consumers' health while enabling market access (Stevenson et al. 2007). Intervention and surveillance for controlling certain diseases in food animals might be driven by public health concerns, although these efforts are not regarded as a serious problem. The welfare and production of farm animals, as well as public health, may be seriously impacted by infectious and other diseases, which often disproportionally affect the most susceptible communities. On-farm food safety inspection is critical to prevent foodborne diseases, build resilient agri-food operations, protect the health of producers and consumers while enabling market access through compliance with relevant international regulations (Stevenson et al. 2007). Inappropriate animal health practices can lead to major productivity and economic losses on the farm, with combined effects across value chains. The most susceptible communities are also most exposed to the consequences of poor animal health because of their limited flexibility (Stringer 2017). Distributing the knowledge of good animal health may be particularly challenging in times of hardship, drought (Agüera et al. 2023; Bahta and Myeki 2021; Khodaverdi et al. 2016; Matope et al. 2020), flooding (Molnár et al. 2021), or natural disaster, as well as in situations of fragility, conflict, and violence. Investments aimed at promoting farming animals' welfare should mitigate risks related to poor conditions affecting animal health (Lekule and Kyvsgaard 2003). Animal health and public health are interrelated and could be improved by installing biosecurity procedures, following good animal husbandry practices, and ensuring the nutritional value of animal-sourced food. Animal health promotes safer and clearer public health systems in many ways.

Protecting animal health attracts public interest. Investments in sustainable farming animals should consider all aspects of health and welfare. Environmental and animal health are also critically linked through biodiversity and ecosystem integrity (World Bank 2018).

Foodborne illnesses can harm national economies by straining healthcare systems, impacting trade and tourism, and stalling the overall socio-economic growth. *Salmonella* (Pomeroy et al. 1989), *Campylobacter* (Carvalho et al. 2010; Gardner and Olson 2018; Kelly et al. 2017), and *Escherichia coli* (Gilbert et al. 2008) are the most important foodborne pathogens that can have severe or fatal outcomes globally. Contaminated foods that contain these pathogens can be of animal origin, for example, eggs, poultry, raw milk, or undercooked meat. The farming animals sector has a shared responsibility for ensuring that the food it produces is safe.

Zoonoses (animal diseases that can be transmitted to humans) (Frazzoli and Mantovani 2010) also impose considerable economic losses in the farming animals sector, which are associated with farming animals' health and welfare, both directly and indirectly. The presence of zoonosis in farming animals can lead to trade barriers, control costs, increased costs because of processing and monitoring to ensure food safety, and losses because of a lack of consumer confidence. The load of zoonoses tends to fall most heavily on developing countries and contributes to an already troubled public health system. The impacts of both the public health and the animal health underscore the connectivity between human health and animal health, as well as the roles of ecosystem alteration and disruption, climate change (Jarmul et al. 2019; Makkar 2018; Malematja et al. 2023), globalization of trade and travel or inadequate biosecurity, among others, which are the main drivers for disease emergence and spread. Some simple practices could safely protect both humans and animals from inter-exchange for harmful microbes and parasites, such as keeping nails trimmed, wearing well-dressed coat, and checking for changes in the skin or fur, such as bald patches, dryness, flakes, or signs of parasites.

Animals can face numerous issues stemming from parasites like fleas, ticks, lice, and mites. The reactions to these pests may vary from mild allergies to severe health problems, including heartworm disease and Lyme disease, which can even be life-threatening. Animals who exhibit unusual behaviors such as shaking their head, scratching their ears, or overly gnawing, licking, or itching their fur should be taken to a veterinarian. Regular health checkups are an important component of caring for animals. They allow monitoring animal welfare (Stevenson et al. 2007). Animals may suffer without showing any obvious signs.

9.5 THE EFFECT OF THE USE OF ANTIBIOTICS AND OTHER ANTIMICROBIAL AGENTS

The widespread overuse and misuse of antibiotics are contributing to the emergence of antibiotic resistance and posing risks to food and environmental safety (Amara et al. 2024; Figure 9.1). Unmanaged risks can have significant consequences on animal health and welfare with consequences on livelihoods, food and nutrition security, public health, biodiversity, and ecosystem services. Safe food can often be taken for granted, and it continues to be a growing public health concern worldwide, especially for susceptible populations (i.e., the very young, old, or immunocompromised).

Antibiotics and other antimicrobial compounds are given to cover the side effects of the crowded negative condition, as well as any weakness or immunocompromised animals; other physiological signs could occur because of overfeeding the animals with excessive amounts of food. In case of poultry farming, studies are conducted to optimize every single step included in the crowded farming process and increase the animal output of the targeted product, farming style, the overall processes, prophylaxis, treatment, vaccination, etc. Sometimes, an antimicrobial formula is delivered orally through water or food intake as a pragmatic solution when animals are kept in larger groups.

Strategies such as disease prevention through immunization, improvement of husbandry practices, and good biosecurity practices that improve animal health and welfare are encouraged as alternatives to antimicrobial use (Amara 2024, 2005, 2022; Amara and Hussain 2006; Amara, Salem-Bekhit, and Alanazi 2013; Amara 2022b). There are a large number of antimicrobial agents that could decrease using antibiotics. For more details, refer to Amara (2023) and Amara et al. (2025).

The nature, specific ecosystems, the air, water, etc., are affected negatively by the antibiotic residues (Amara et al. 2024). Other factors that affect both humans and animals could not be ignored, which include factors related to politic, economy, health, social, welfare, and time and the agricultural areas that are used to provide everyone with food.

9.6 ANIMALS' HYGIENE AND ITS RELATED ISSUES

Animals' hygiene refers to the practices used to protect animals and human health and the environment where animals are raised, bred, and given care. Farmers' hygiene cannot be maintained without acknowledging the interdependence of people, animals, and the environment (Tuma 2024). As a result, defense against the possibility of pathogen transmission is offered. Farmers, workers, and veterinarians who work with animals and the environment need to rank their hygiene in the farm, different farming related areas, and clinics, in order to avoid disturbing facility hygiene and the overall system hygiene (Tuma 2024). In order to maintain good personal hygiene, people who produce food from farming animals should dress appropriately for their jobs, wash their hands frequently, and keep their workspace well ordered. To maintain animals' hygiene, both the animals' environment and the person in contact with animals are in need to be cleaned and disinfected (Tuma 2024). Every surface in the environment needs to be cleaned, disinfected, and inspected often (Hussein and Amara 2006). The technique's instruments need to be routinely inspected, cleaned, and disinfected. Not only should the facility be kept clean and sterilized but also each single contact must be controlled. People's and cars' access and departure times should be controlled. When one's work clothes become contaminated, they must be washed properly. Hand hygiene needs to be given more attention (Tuma 2024). They should start by washing their hands with soap and water, followed by the use of hand sanitizers, to maintain cleanliness and disinfection both in the environment and for individuals. Keeping animals clean and covered is important and vital for their overall health and welfare. Effective sanitation practices are necessary to provide a suitable living environment for animals and to prevent various diseases. Understanding and applying the principles of animals' hygiene are crucial for responsible animal care and ethical animal proprietorship whether for household animals or farming animals. Regular brushing helps animals maintain a healthy coat and good circulation, preventing fur matting.

Farming animals like cattle and horses should be cleaned regularly to avoid infections and promote skin health. Covered sessions also offer animal owners a chance to strengthen their contact with their animals, improving their mental and emotional welfare. Animals' hygiene involves routine health checks, clean-living conditions, coating, and proper nutrition. By applying these aspects, farmers and animal owners can ensure the welfare of their animals. These practices not only enhance physical health but also positively impact mental and emotional well-being, leading to a happy and satisfying natural life. It is important to uphold high standards of animals' hygiene to ensure proper and happy animals' welfare (Tuma 2024).

9.7 ANIMAL–HUMAN INTERACTIONS

A proper diet along with regular physical activity is crucial for the animals. Consistent exercise enhances animal mood while keeping them strong and fit and enhances their acceptance of the food. Factors such as age, breed, size, and general health can affect how much exercise an animal needs. An animal's health greatly depends on having access to clean drinking water. It is vital to ensure that the animal has daily access to fresh, drinkable water. Impure water can cause vomiting, diarrhea, and other digestive issues. Much like humans, animals require daily attention to their dental health. Ignoring animals' dental care can result in serious heart, liver, and kidney problems, as well as tooth loss and gum disease as they age. Animals should not be kept in excessively hot or cold environments. Exposure to frigid, dry air, rain, or snow can lead to itchy skin and cracked paws in animals. When animals are subjected to chilly temperatures without appropriate protection, there's a risk of hypothermia. During extreme weather, either warm or cold, ensure that animals stay inside. Changes in eating patterns, sleep routines, restlessness, and mood might indicate a hidden medical issue. By keeping the veterinarian informed about any variations, one could assist in identifying potential problems before they worsen. Animal showering will guarantee many positive health and mental conditions. Human companionship of animals is part of animal welfare (Verga and and Michelazzi 2009).

9.8 CONCLUSION

Animal farms come in a variety of forms. Under each form, there are many variations based on many factors such as the animals, humans, locations, environment, economic power, and market size. But there is usually a general role that can be followed to guarantee safe food, good health, comfortable living, safe environment, perfect products with high quality, and safe practices for both humans and animals. Other issues are also included. The most successful, popular style is the family farming where the concept of "we all are happy," could be observed. Its power is in its simplicity. It can be used as a tool to analyze what was changed and what could be repaired, replaced, and enhanced. In fact, humans are at the end of the farming animal product chain. Humans will finally be affected by the overall process through consuming its output represented in the different farm products. This chapter discusses a countable number of problems concerning the animal food, animal feeding, feeding style, and process; and considerable number of factors that could transfer pathogens, toxins, chemical compounds to the food or to a person who finally will transfer it to the other foods or to the animal food, because we should apply "one health" strategy to be able to control the animal farming process for the best animal welfare.

REFERENCES

Abdisa, T., and T. Dilbato. 2024. "Toxic plants and their impact on livestock health and economic losses: A comprehensive review." *J Toxicol* 2024:9857933. doi:10.1155/jt/9857933.

Aguayo-Ulloa, L. A., G. C. Miranda-de la Lama, M. Pascual-Alonso, J. L. Olleta, M. Villarroel, C. Sañudo, and G. A. María. 2014. "Effect of enriched housing on welfare, production performance and meat quality in finishing lambs: The use of feeder ramps." *Meat Sci* 97 (1):42–8. doi:10.1016/j.meatsci.2014.01.001.

Agüera, E. I., L. Requena, M. B. García-Moreno, M. A. Pérez-Priego, and F. Requena. 2023. "Promotion of pregnant merino ewes' welfare with the introduction of a drought- and high-temperature-resistant cereal into their diet: Analysis of tritordeum meadow." *Animals* 13 (19). doi:10.3390/ani13193155.

Ajila, C. M., Brar S. K., Verma M., Tyagi R. D., Godbout S., and J. R. and Valéro. 2012. "Bio-processing of agro-byproducts to animal feed." *Crit Rev Biotechnol* 32 (4):382–400. doi:10.3109/07388551.2012.659172.

Al-Mutairi, M. H., K. Alzahrani, A. T. Dabiah, and H. S. Kassem. 2023. "Adoption of on-farm feed safety practices among livestock farmers: Evidence from Saudi Arabia." *Heliyon* 9 (12):e22838. doi:10.1016/j.heliyon.2023.e22838.

Alfonso, S., E. Mente, E. Fiocchi, A. Manfrin, A. Dimitroglou, L. Papaharisis, D. Barkas, L. Toomey, M. Boscarato, C. Losasso, A. Peruzzo, A. Stefani, W. Zupa, M. T. Spedicato, I. Nengas, G. Lembo, and P. Carbonara. 2023. "Growth performance, gut microbiota composition, health and welfare of European sea bass (Dicentrarchus labrax) fed an environmentally and economically sustainable low marine protein diet in sea cages." *Sci Rep* 13 (1):21269. doi:10.1038/s41598-023-48533-3.

Amara, A. A. 2005. "New strategy for controlling resistant strain of Pseudomonas aeruginosa." In: *Proceeding of Tagungsband zum 2. Gemeinsamen kongress der DGHM und VAAM Göttingen*, Göttingen, *Germany,* 25–28 September 2005 (*BioSpectrum*:74).

Amara, A. A. 2022a. "The role of divalent cations in antibiotic sensitivity: A molecular aspect." In: V. K. Gupta, S. D. Sarker, M. Sharma, M. E. Pirovani, Z. Usmani, and C. Jayabaskaran (Eds.), *Biomolecules from Natural Sources: Advances and Applications* (pp. 252–277). Wiley.

Amara, A. A. 2022b. "The role of divalent cations in antibiotic sensitivity." In: V. K. Gupta, S. D. Sarker, M. Sharma, M. E. Pirovani, Z. Usmani, and C. Jayabaskaran (Eds.), *Biomolecules from Natural Sources*. Wiley.

Amara, A. A. 2023. "Fermented food microbes and their preservative byproducts." In: A. Gupta and R. Prasad (Eds.), *Antimicrobials in Food Science and Technology* (Original edition). CRC Press.

Amara, A. A. 2024. "Phags and their role in naturaland directed microbial control." In: S. Varjani, I. Haq, A. Pandey, V. K. Gupta, and X. T. Bui (Eds.), *Waste Management in Climate Change and Sustainability Perspectives*. CRC Press.

Amara, A. A. A. F. 2013. "The inevitability of balanced lives: Genes and, foods in action and, interactions." *IIOAB J* 4 (2):1–27.

Amara, A. A., and A. Gupta. 2025. "An open traditional poultry production, an increasing interest toward healthier animals with less drug additives." In: A. Gupta and R. Prasad (Eds.), *Antimicrobials in Animal Husbandry*. CRC Press.

Amara, A. A., and A. Shibl. 2015. "A hypothesis about the variation within β-lactamases: An epigenetic-like mechanism." *Inst Integr Omics Appl Biotechnol J* 6 (1):1–11.

Amara, A. A., M. M. Salem-Bekhit, and F. K. Alanazi. 2013. "Sponge-like: A new protocol for preparing bacterial ghosts." *Sci World J* 2013. doi:10.1155/2013/545741.

Amara, A. A., M. M. Salem-Bekhit, K. M. A. Zoheir, A. Gupta, and M. M. M. Ahmed. 2025. "The antimicrobial resistance in agriculture and its role in human and animal health." In: A. Gupta and R. Prasad (Eds.), *Antimicrobials in Animal Husbandry*. Taylor & Francis.

Amara, A. A., N. Abd El-Baky, M. M. Salem-Bekhit, Khairy M. A. Zoheir, and M. M. M. Ahmed. 2024. Antibiotic resistance from theory to facts. In: A. Gupta and R. Prasad Gupta (Eds.), *Antimicrobial Resistance in Agriculture and its Consequences*. CRC Press.

Amara, A. A., and Z. M. Hussain. 2006. "Offline controlling of *Pseudomonas aeruginosa* resistant to protein inhibitor antibiotics using combination of EDTA and Na-citrate or disinfectant(s)." *Microbiology* 33 (5):95–100.

Arieli, A., J. E. Vallimont, Y. Aharoni, and G. A. Varga. 2001. "Monensin and growth hormone effects on glucose metabolism in the prepartum cow." *J Dairy Sci* 84 (12):2770–6. doi:10.3168/jds.S0022-0302(01)74731-5.

Ashley, K., S. Wilson, J. R. Young, H. P. Chan, S. Vitou, S. Suon, P. A. Windsor, and R. D. Bush. 2018. "Drivers, challenges and opportunities of forage technology adoption by smallholder cattle households in Cambodia." *Trop Anim Health Prod* 50 (1):63–73. doi:10.1007/s11250-017-1400-y.

Attia, Y. A., E. A. Ashour, S. A. Nagadi, M. R. Farag, F. Bovera, and M. Alagawany. 2023. "Rice bran as an alternative feedstuff in broiler nutrition and impact of Liposorb® and vitamin E-Se on sustainability of performance, carcass traits, blood biochemistry, and antioxidant indices." *Vet Sci* 10 (4). doi:10.3390/vetsci10040299.

Bahta, Y. T., and V. A. Myeki. 2021. "Adaptation, coping strategies and resilience of agricultural drought in South Africa: Implication for the sustainability of livestock sector." *Heliyon* 7 (11):e08280. doi:10.1016/j.heliyon.2021.e08280.

Baldin, M., T. Breunig, R. Cue, A. De Vries, M. Doornink, J. Drevenak, R. Fourdraine, R. George, R. Goodling, R. Greenfield, M. W. Jorgensen, A. Lenkaitis, D. Reinemann, A. Saha, C. Sankaraiah, S. Shahinfar, C. Siberski, K. M. Wade, F. Zhang, L. Fadul-Pacheco, S. Wangen, T. E. da Silva, and V. E. Cabrera. 2021. "Integrated Decision Support Systems (IDSS) for dairy farming: A discussion on how to improve their sustained adoption." *Animals* 11 (7). doi:10.3390/ani11072025.

Brice, J., G. Cusworth, J. Lorimer, and T. Garnett. 2022. "Immaterial animals and financialized forests: Asset manager capitalism, ESG integration and the politics of livestock." *Environ Plan A* 54 (8):1551–68. doi:10.1177/0308518x221121132.

Broom, D. M. 2010. "Animal welfare: An aspect of care, sustainability, and food quality required by the public." *J Vet Med Educ* 37 (1):83–8. doi:10.3138/jvme.37.1.83.

Broom, D. M., F. A. Galindo, and E. Murgueitio. 2013. "Sustainable, efficient livestock production with high biodiversity and good welfare for animals." *Proc Biol Sci* 280 (1771):20132025. doi:10.1098/rspb.2013.2025.

Carvalho, C. M., B. W. Gannon, D. E. Halfhide, S. B. Santos, C. M. Hayes, J. M. Roe, and J. Azeredo. 2010. "The in vivo efficacy of two administration routes of a phage cocktail to reduce numbers of Campylobacter coli and Campylobacter jejuni in chickens." *BMC Microbiol* 10:232. doi:10.1186/1471-2180-10-232.

Colpoys, J., D. Van Sambeek, C. Bruns, A. Johnson, J. Dekkers, F. Dunshea, and N. Gabler. 2019. "Responsiveness of swine divergently selected for feed efficiency to exogenous adrenocorticotropic hormone and glucose challenges." *Domest Anim Endocrinol* 68:32–38. doi:10.1016/j.domaniend.2018.12.007.

Cooke, A. S., H. Machekano, L. C. Gwiriri, J. H. I. Tinsley, G. M. Silva, C. Nyamukondiwa, A. Safalaoh, E. R. Morgan, and M. R. F. Lee. 2025. "The nutritional feed gap: Seasonal variations in ruminant nutrition and knowledge gaps in relation to food security in Southern Africa." *Food Secur* 17 (1):73–100. doi:10.1007/s12571-024-01509-1.

Cornish, A., D. Raubenheimer, and P. McGreevy. 2016. "What we know about the public's level of concern for farm animal welfare in food production in developed countries." *Animals* 6 (11). doi:10.3390/ani6110074.

Crovato, S., A. Pinto, G. Di Martino, G. Mascarello, V. Rizzoli, S. Marcolin, and L. Ravarotto. 2022. "Purchasing habits, sustainability perceptions, and welfare concerns of italian consumers regarding rabbit meat." *Foods* 11 (9). doi:10.3390/foods11091205.

Dartt, B. A., J. W. Lloyd, B. R. Radke, J. R. Black, and J. B. Kaneene. 1999. "A comparison of profitability and economic efficiencies between management-intensive grazing and conventionally managed dairies in Michigan." *J Dairy Sci* 82 (11):2412–20. doi:10.3168/jds.S0022-0302(99)75492-5.

Dawood, M. A. O., S. E. Abdo, S. El-Kassas, K. El-Naggar, R. A. Al Wakeel, E. M. Moustafa, and S. Abou Asa. 2024. "Chicken egg lysozyme enhanced the growth performance, feed utilization, upregulated immune-related genes, and mitigated the impacts of Aeromonas hydrophila infection in Nile tilapia (Oreochromisniloticus)." *Fish Shellfish Immunol* 146:109377. doi:10.1016/j.fsi.2024.109377.

Decuypere, E., V. Bruggeman, N. Everaert, Y. Li, R. Boonen, J. De Tavernier, S. Janssens, and N. Buys. 2010. "The broiler breeder paradox: Ethical, genetic and physiological perspectives, and suggestions for solutions." *Br Poult Sci* 51 (5):569–79. doi:10.1080/00071668.2010.519121.

Delvendahl, N., B. A. Rumpold, and N. Langen. 2022. "Edible insects as food-insect welfare and ethical aspects from a consumer perspective." *Insects* 13 (2). doi:10.3390/insects13020121.

Desbois, A. P., M. Garza, M. Eltholth, Y. M. Hegazy, A. Mateus, A. Adams, D. C. Little, E. Høg, C. V. Mohan, S. E. Ali, and L. A. Brunton. 2021. "Systems-thinking approach to identify and assess feasibility of potential interventions to reduce antibiotic use in tilapia farming in Egypt." *Aquaculture* 540:736735. doi:10.1016/j.aquaculture.2021.736735.

Dias, I., M. Laranjo, M. E. Potes, A. C. Agulheiro-Santos, S. Ricardo-Rodrigues, A. R. Fialho, J. Véstia, M. J. Fraqueza, M. Oliveira, and M. Elias. 2020. "Autochthonous starter cultures are able to reduce biogenic amines in a traditional Portuguese smoked fermented sausage." *Microorganisms* 8 (5). doi:10.3390/microorganisms8050686.

Diaz Vicuna, E., K. Srikanthithasan, R. Odore, S. Massaglia, V. M. Merlino, A. Giorgino, L. Ozella, A. Schiavone, F. R. Massacci, J. Mota-Gutierrez, and C. Forte. 2024. "Influence of age, gender, and willingness to adopt former foodstuffs on the perception of Italian farm animal veterinarians." *Front Vet Sci* 11:1396807. doi:10.3389/fvets.2024.1396807.

Dong, L., L. Zhao, B. Li, Y. Gao, T. Yan, P. Lund, Z. Liu, and Q. Diao. 2023. "Dietary supplementation with xylooligosaccharides and exogenous enzyme improves milk production, energy utilization efficiency and reduces enteric methane emissions of Jersey cows." *J Anim Sci Biotechnol* 14 (1):71. doi:10.1186/s40104-023-00873-w.

Ebrahimi, M., N. H. Daeman, C. M. Chong, A. Karami, V. Kumar, S. H. Hoseinifar, and N. Romano. 2017. "Comparing the effects of different dietary organic acids on the growth, intestinal short-chain fatty acids, and liver histopathology of red hybrid tilapia (Oreochromis sp.) and potential use of these as preservatives." *Fish Physiol Biochem* 43 (4):1195–207. doi:10.1007/s10695-017-0365-0.

El-Sabrout, K., A. Khalifah, and B. Mishra. 2023. "Application of botanical products as nutraceutical feed additives for improving poultry health and production." *Vet World* 16 (2):369–79. doi:10.14202/vetworld.2023.369-379.

Erensoy, K., M. Sarıca, M. Noubandiguim, M. Dur, and R. Aslan. 2021. "Effect of light intensity and stocking density on the performance, egg quality, and feather condition of laying hens reared in a battery cage system over the first laying period." *Trop Anim Health Prod* 53 (2):320. doi:10.1007/s11250-021-02765-5.

Erickson, P. S., J. L. Anderson, K. F. Kalscheur, G. J. Lascano, M. S. Akins, and A. J. Heinrichs. 2020. "Symposium review: Strategies to improve the efficiency and profitability of heifer raising." *J Dairy Sci* 103 (6):5700–08. doi:10.3168/jds.2019-17419.

Eugenio, F. A., J. van Milgen, J. Duperray, R. Sergheraert, and N. Le Floc'h. 2022. "Feeding intact proteins, peptides, or free amino acids to monogastric farm animals." *Amino Acids* 54 (2):157–68. doi:10.1007/s00726-021-03118-0.

Fatumah, N., S. Mohammed, N. Ashraf, K. Abasi, and N. Shadia. 2023. "Adoption of novel climate-smart farming systems for enhanced carbon stock and carbon dioxide equivalent emission reduction in cattle corridor areas of Uganda." *Heliyon* 9 (3):e14114. doi:10.1016/j.heliyon.2023.e14114.

Federico, J. L., and R. E. Banks. 2020. "PET PODs: A tool to assist with good welfare provisions during emergent conditions." *J Appl Anim Welf Sci* 23 (3):253–64. doi:10.1080/10888705.2019.1619179.

Fiorilla, E., I. Ferrocino, M. Gariglio, F. Gai, V. Zambotto, L. Ozella, I. Franciosa, M. Giribaldi, S. Antoniazzi, F. Raspa, E. E. Cappone, D. Fabrikov, S. Pozzo, V. Bongiorno, A. Calissano, D. Ippolito, S. Bergagna, K. Srikanthithasan, C. Forte, M. T. Capucchio, and A. Schiavone. 2024. "Black soldier fly larvae: A one health approach to investigate gut, and organ health and meat quality response in slow-growing chickens." *BMC Vet Res* 20 (1):580. doi:10.1186/s12917-024-04394-0.

Flachowsky, G. 2011. "Carbon-footprints for food of animal origin, reduction potentials and research need." *J Appl Anim Res* 39 (1):2–14. doi:10.1080/09712119.2011.570047.

Fraser, D. 2001. "Farm animal production: Changing agriculture in a changing culture." *J Appl Anim Welf Sci* 4 (3):175–90. doi:10.1207/S15327604JAWS0403_02.

Frazzoli, C., and A. Mantovani. 2010. "Toxicants exposures as novel zoonoses: Reflections on sustainable development, food safety and veterinary public health." *Zoonoses Public Health* 57 (7–8):e136–42. doi:10.1111/j.1863-2378.2009.01309.x.

Fu, H., H. Chen, Q. Ma, B. Chen, F. Wang, and L. Wu. 2023. "Planting and mowing cover crops as livestock feed to synergistically optimize soil properties, economic profit, and environmental burden on pear orchards in the Yangtze River Basin." *J Sci Food Agric* 103 (13):6680–8. doi:10.1002/jsfa.12763.

Gardner, S. P., and J. W. Olson. 2018. "Interaction of copper toxicity and oxidative stress in campylobacter jejuni." *J Bacteriol* 200 (21). doi:10.1128/jb.00208-18.

Gasco, L., I. Biasato, S. Dabbou, A. Schiavone, and F. Gai. 2019. "Animals fed insect-based diets: State-of-the-art on digestibility, performance and product quality." *Animals* 9 (4). doi:10.3390/ani9040170.

Gilbert, R. A., S. E. Denman, J. Padmanabha, N. Fegan, D. Al Ajmi, and C. S. McSweeney. 2008. "Effect of diet on the concentration of complex Shiga toxin-producing *Escherichia coli* and EHEC virulence genes in bovine faeces, hide and carcass." *Int J Food Microbiol* 121 (2):208–16. doi:10.1016/j.ijfoodmicro.2007.11.019.

Gill, M. 1999. "Meat production in developing countries." *Proc Nutr Soc* 58 (2):371–6. doi:10.1017/s002966519900049x.

Goumeida, K., D. Rahmani, J. Le Cour Grandmaison, and J. M. Gil Roig. 2024. "Citizens' willingness to support farmers' adoption of low crude protein diet in pig production." *Heliyon* 10 (7):e28211. doi:10.1016/j.heliyon.2024.e28211.

Graham-Thiers, P., and K. Bowen. 2024. "Timing of feeding a protein supplement on nitrogen balance and plasma amino acids during exercise recovery in horses." *J Anim Physiol Anim Nutr* 108 (6):1723–33. doi:10.1111/jpn.14012.

Greenwood, P. L. 2021. "Review: An overview of beef production from pasture and feedlot globally, as demand for beef and the need for sustainable practices increase." *Animal* 15 Suppl 1:100295. doi:10.1016/j.animal.2021.100295.

Hadfield, J. A., B. L. Waldron, S. C. Isom, R. Feuz, R. Larsen, J. E. Creech, M. F. Rose, J. Long, M. D. Peel, R. L. Miller, K. A. Rood, A. Young, R. Stott, A. Sweat, and K. J. Thornton. 2021. "The effects of organic grass and grass-birdsfoot trefoil pastures on Jersey heifer development: Heifer growth, performance, and economic impact." *J Dairy Sci* 104 (10):10863–78. doi:10.3168/jds.2020-19524.

Halmemies-Beauchet-Filleau, A., M. Rinne, M. Lamminen, C. Mapato, T. Ampapon, M. Wanapat, and A. Vanhatalo. 2018. "Review: Alternative and novel feeds for ruminants: Nutritive value, product quality and environmental aspects." *Animal* 12 (s2):s295–s309. doi:10.1017/s1751731118002252.

Hanson, G. D., L. C. Cunningham, M. J. Morehart, and R. L. Parsons. 1998. "Profitability of moderate intensive grazing of diary cows in the northeast." *J Dairy Sci* 81 (3):821–9. doi:10.3168/jds.s0022-0302(98)75640-1.

Heinrichs, A. J., C. M. Jones, P. S. Erickson, H. Chester-Jones, and J. L. Anderson. 2020. "Symposium review: Colostrum management and calf nutrition for profitable and sustainable dairy farms." *J Dairy Sci* 103 (6):5694–99. doi:10.3168/jds.2019-17408.

Herrero, M., P. K. Thornton, A. M. Notenbaert, S. Wood, S. Msangi, H. A. Freeman, D. Bossio, J. Dixon, M. Peters, J. van de Steeg, J. Lynam, P. Parthasarathy Rao, S. Macmillan, B. Gerard, J. McDermott, C. Seré, and M. Rosegrant. 2010. "Smart investments in sustainable food production: Revisiting mixed crop-livestock systems." *Science* 327 (5967):822–5. doi:10.1126/science.1183725.

Hocquette, J.-F., R.R. Ian, P. Sophie, M. Françoise, D. Geraldine, and N. D. Scollan. 2005. "The future trends for research on quality and safety of animal products." *Italian J Anim Sci* 4 (sup3):49–72. doi:10.4081/ijas.2005.3s.49.

Huang, J., X. F. Tong, Z. W. Yu, Y. P. Hu, L. Zhang, Y. Liu, and Z. X. Zhou. 2020. "Dietary supplementation of total flavonoids from Rhizoma Drynariae improves bone health in older caged laying hens." *Poult Sci* 99 (10):5047–54. doi:10.1016/j.psj.2020.06.057.

Hussein, M. Z., and A. A. Amara. 2006. "Case-by-case study using antibiotic-EDTA combination to control pseudomonas aeruginosa." *Pakist J Pharmaceut Sci* 19 (3):236–43.

Jarmul, S., Z. Liew, A. Haines, and P. Scheelbeek. 2019. "Climate change mitigation in food systems: The environmental and health impacts of shifting towards sustainable diets, a systematic review protocol." *Wellcome Open Res* 4:205. doi:10.12688/wellcomeopenres.15618.1.

Karlsson, J., M. Lindberg, M. Åkerlind, and K. Holtenius. 2020. "Whole-lactation feed intake, milk yield, and energy balance of Holstein and Swedish Red dairy cows fed grass-clover silage and 2 levels of byproduct-based concentrate." *J Dairy Sci* 103 (10):8922–37. doi:10.3168/jds.2020-18204.

Kelly, C., O. Gundogdu, G. Pircalabioru, A. Cean, P. Scates, M. Linton, L. Pinkerton, E. Magowan, L. Stef, E. Simiz, I. Pet, S. Stewart, R. Stabler, B. Wren, N. Dorrell, and N. Corcionivoschi. 2017. "The in vitro and in vivo effect of carvacrol in preventing campylobacter infection, colonization and in improving productivity of chicken broilers." *Foodborne Pathog Dis* 14 (6):341–9. doi:10.1089/fpd.2016.2265.

Khodaverdi, H., T. Fowles, E. Bick, and C. Nansen. 2016. "Does drought increase the risk of insects developing behavioral resistance to systemic insecticides?" *J Econ Entomol* 109 (5):2027–31. doi:10.1093/jee/tow188.

Kijlstra, A., and I. A. J. M. Eijck. 2006. "Animal health in organic livestock production systems: A review." *NJAS: Wageningen J Life Sci* 54 (1):77–94. doi:10.1016/S1573-5214(06)80005-9.

Laso, J., M. Margallo, I. García-Herrero, P. Fullana, A. Bala, C. Gazulla, A. Polettini, R. Kahhat, I. Vázquez-Rowe, A. Irabien, and R. Aldaco. 2018. "Combined application of Life Cycle Assessment and linear programming to evaluate food waste-to-food strategies: Seeking for answers in the nexus approach." *Waste Manag* 80:186–97. doi:10.1016/j.wasman.2018.09.009.

Lekule, F. P., and N. C. Kyvsgaard. 2003. "Improving pig husbandry in tropical resource-poor communities and its potential to reduce risk of porcine cysticercosis." *Acta Trop* 87 (1):111–7. doi:10.1016/s0001-706x(03)00026-3.

Leroy, F., F. Abraini, T. Beal, P. Dominguez-Salas, P. Gregorini, P. Manzano, J. Rowntree, and S. van Vliet. 2022. "Animal board invited review: Animal source foods in healthy, sustainable, and ethical diets - An argument against drastic limitation of livestock in the food system." *Animal* 16 (3):100457. doi:10.1016/j.animal.2022.100457.

Li, H., X. Wen, R. Alphin, Z. Zhu, and Z. Zhou. 2017. "Effects of two different broiler flooring systems on production performances, welfare, and environment under commercial production conditions." *Poult Sci* 96 (5):1108–19. doi:10.3382/ps/pew440.

Lingbeck, J. M., P. Cordero, C. A. O'Bryan, M. G. Johnson, S. C. Ricke, and P. G. Crandall. 2014. "Functionality of liquid smoke as an all-natural antimicrobial in food preservation." *Meat Sci* 97 (2):197–206. doi:10.1016/j.meatsci.2014.02.003.

Liu, B., Y. Zhou, X. Xia, C. Wang, H. Wei, and J. Peng. 2020. "Effects of dietary lysine levels on production performance and milk composition of high-producing sows during lactation." *Animals* 10 (11). doi:10.3390/ani10111947.

Liu, N., Q. Zhu, X. Zeng, B. Yang, M. Liang, L. Deng, L. He, C. Liang, R. Zhang, and J. Zhou. 2018. "Optimization of processing conditions of Chinese smoke-cured bacon (Larou) with a new natural coating solution during storage period." *Korean J Food Sci Anim Resour* 38 (3):636–52. doi:10.5851/kosfa.2018.38.3.636.

Luise, D., P. Bosi, L. Raff, L. Amatucci, S. Virdis, and P. Trevisi. 2022. "Bacillus spp. Probiotic Strains as a potential tool for limiting the use of antibiotics, and improving the growth and health of pigs and chickens." *Front Microbiol* 13:801827. doi:10.3389/fmicb.2022.801827.

Luo, S., Y. Li, S. Chen, D. Zhang, and S. S. Renner. 2011. "Gelechiidae moths are capable of chemically dissolving the pollen of their host plants: First documented sporopollenin breakdown by an animal." *PLoS One* 6 (4):e19219. doi:10.1371/journal.pone.0019219.

Makkar, H. P. S. 2018. "Review: Feed demand landscape and implications of food-not feed strategy for food security and climate change." *Animal* 12 (8):1744–54. doi:10.1017/s175173111700324x.

Maleko, D., W. T. Ng, G. Msalya, A. Mwilawa, L. Pasape, and K. Mtei. 2018. "Seasonal variations in the availability of fodder resources and practices of dairy cattle feeding among the smallholder farmers in Western Usambara Highlands, Tanzania." *Trop Anim Health Prod* 50 (7):1653–64. doi:10.1007/s11250-018-1609-4.

Malematja, E., N. A. Sebola, T. G. Manyelo, S. D. Kolobe, and M. Mabelebele. 2023. "Copping out of novel feeds: HOW climate change pledgers and food summits overlooked insect protein." *Heliyon* 9 (12):e22773. doi:10.1016/j.heliyon.2023.e22773.

Mantovani, A., G. Aquilina, F. Cubadda, and F. Marcon. 2022. "Risk-benefit assessment of feed additives in the one health perspective." *Front Nutr* 9:843124. doi:10.3389/fnut.2022.843124.

Marques, J. G. O., O. Silva R. de, L. G. Barioni, J. A. J. Hall, L. O. Tedeschi, and D. Moran. 2020. "An improved algorithm for solving profit-maximizing cattle diet problems." *Animal* 14 (S2):s257–66. doi:10.1017/s1751731120001433.

Martelli, G. 2009. "Consumers' perception of farm animal welfare: An Italian and European perspective." *Italian J Anim Sci* 8 (sup1):31–41. doi:10.4081/ijas.2009.s1.31.

Matope, A., T. J. Zindove, M. Dhliwayo, and M. Chimonyo. 2020. "Mitigating the effects of drought on cattle production in communal rangelands of Zimbabwe." *Trop Anim Health Prod* 52 (1):321–30. doi:10.1007/s11250-019-02020-y.

Meijer, N., L. W. D. Van Raamsdonk, E. W. J. Gerrits, and M. J. Appel. 2023. "The use of animal by-products in a circular bioeconomy: Time for a TSE road map 3?" *Heliyon* 9 (3):e14021. doi:10.1016/j.heliyon.2023.e14021.

Melchior, D., B. Sève, and N. Le Floc'h. 2004. "Chronic lung inflammation affects plasma amino acid concentrations in pigs." *J Anim Sci* 82 (4):1091–9. doi:10.2527/2004.8241091x.

Mohd Azmi, A. F., F. Mat Amin, H. Ahmad, N. Mohd Nor, G. Y. Meng, M. Zamri Saad, M. Z. Abu Bakar, P. Abdullah, A. Irawan, A. Jayanegara, and H. Abu Hassim. 2021. "Effects of bypass fat on buffalo carcass characteristics, meat nutrient contents and profitability." *Animals* 11 (11). doi:10.3390/ani11113042.

Molnár, Z., K. Szabados, A. Kiš, J. Marinkov, L. Demeter, M. Biró, K. Öllerer, K. Katona, M. Đapić, R. Perić, V. Ulicsni, and D. Babai. 2021. "Preserving for the future the—once widespread but now vanishing—knowledge on traditional pig grazing in forests and marshes (Sava-Bosut floodplain, Serbia)." *J Ethnobiol Ethnomed* 17 (1):56. doi:10.1186/s13002-021-00482-9.

Muir, W. I., Y. Akter, K. Bruerton, and P. J. Groves. 2022. "An evaluation of bird weight and diet nutrient density during early lay on ISA Brown performance, egg quality, bone characteristics, and liver health at 50 weeks of age." *Poult Sci* 101 (5):101765. doi:10.1016/j.psj.2022.101765.

Naab, F. Z., D. Coles, E. Goddard, and L. J. Frewer. 2021. "Public perceptions regarding genomic technologies applied to breeding farm animals: A qualitative study." *BioTech* 10 (4). doi:10.3390/biotech10040028.

Nair, M. R. R., V. Sejian, M. V. Silpa, V. F. C. Fonsêca, C. C. de Melo Costa, C. Devaraj, G. Krishnan, M. Bagath, P. O. Nameer, and R. Bhatta. 2021. "Goat as the ideal climate-resilient animal model in tropical environment: Revisiting advantages over other livestock species." *Int J Biometeorol* 65 (12):2229–40. doi:10.1007/s00484-021-02179-w.

Nalubwama, S. M., A. Mugisha, and M. Vaarst. 2011. "Organic livestock production in Uganda: Potentials, challenges and prospects." *Trop Anim Health Prod* 43 (4):749–57. doi:10.1007/s11250-011-9780-x.

Nathanson, N., J. Wilesmith, and C. Griot. 1997. "Bovine spongiform encephalopathy (BSE): Causes and consequences of a common source epidemic." *Am J Epidemiol* 145 (11):959–69. doi:10.1093/oxfordjournals.aje.a009064.

Nikmaram, N., and K. A. Rosentrater. 2019. "Overview of some recent advances in improving water and energy efficiencies in food processing factories." *Front Nutr* 6:20. doi:10.3389/fnut.2019.00020.

Ogbuewu, I. P., and C. A. Mbajiorgu. 2023. "Potentials of dietary zinc supplementation in improving growth performance, health status, and meat quality of broiler chickens." *Biol Trace Elem Res* 201 (3):1418–31. doi:10.1007/s12011-022-03223-5.

Opoola, O., F. Shumbusho, D. Hambrook, S. Thomson, H. Dai, M. G. G. Chagunda, J. L. Capper, D. Moran, R. Mrode, and A. Djikeng. 2022. "From a documented past of the Jersey breed in Africa to a profit index linked future." *Front Genet* 13:881445. doi:10.3389/fgene.2022.881445.

Øverland, M., L. T. Mydland, and A. Skrede. 2019. "Marine macroalgae as sources of protein and bioactive compounds in feed for monogastric animals." *J Sci Food Agric* 99 (1):13–24. doi:10.1002/jsfa.9143.

Pahlow, M., P. R. van Oel, M. M. Mekonnen, and A. Y. Hoekstra. 2015. "Increasing pressure on freshwater resources due to terrestrial feed ingredients for aquaculture production." *Sci Total Environ* 536:847–57. doi:10.1016/j.scitotenv.2015.07.124.

Paiva, R. C., C. A. Moura, P. Thomas, B. Haberl, L. Greiner, C. J. Rademacher, Apsp Silva, G. Trevisan, D. C. L. Linhares, and G. S. Silva. 2023. "Risk factors associated with sow mortality in breeding herds under one production system in the Midwestern United States." *Prev Vet Med* 213:105883. doi:10.1016/j.prevetmed.2023.105883.

Phillips, C. J. C. 2024. "Farm animal welfare-from the farmers' perspective." *Animals* 14 (5). doi:10.3390/ani14050671.

Pomeroy, B. S., K. V. Nagaraja, L. T. Ausherman, I. L. Peterson, and K. A. Friendshuh. 1989. "Studies on feasibility of producing Salmonella-free turkeys." *Avian Dis* 33 (1):1–7.

Poosuwan, K., C. Bunchasak, and C. Kaewtapee. 2010. "Long-term feeding effects of dietary protein levels on egg production, immunocompetence and plasma amino acids of laying hens in subtropical condition." *J Anim Physiol Anim Nutr* 94 (2):186–95. doi:10.1111/j.1439-0396.2008.00898.x.

Preston, T. R. 2009. "Environmentally sustainable production of food, feed and fuel from natural resources in the tropics." *Trop Anim Health Prod* 41 (7):1071–80. doi:10.1007/s11250-009-9453-1.

Qiao, G., X. Li, J. Li, M. Zhang, Y. Shen, Z. Zhao, Y. Zhang, Z. Qi, P. Chen, Y. Sun, P. Cang, P. Liu, E. Wangkahart, and Z. Wang. 2022. "An eco-friendly conversion of aquaculture suspended solid wastes into high-quality fish food by improving poly-β-hydroxybutyrate production." *Front Physiol* 13:797625. doi:10.3389/fphys.2022.797625.

Randazzo, B., M. Zarantoniello, G. Gioacchini, E. Giorgini, C. Truzzi, V. Notarstefano, G. Cardinaletti, K. T. Huyen, O. Carnevali, and I. Olivotto. 2020. "Can insect-based diets affect zebrafish (*Danio rerio*) reproduction? A multidisciplinary study." *Zebrafish* 17 (5):287–304. doi:10.1089/zeb.2020.1891.

Rashid, Z., Z. A. Mirani, S. Zehra, S. M. H. Gilani, A. Ashraf, A. Azhar, K. A. Al-Ghanim, F. Al-Misned, N. Al-Mulahim, S. Mahboob, and S. Galani. 2020. "Enhanced modulation of gut microbial dynamics affecting body weight in birds triggered by natural growth promoters administered in conventional feed." *Saudi J Biol Sci* 27 (10):2747–55. doi:10.1016/j.sjbs.2020.06.027.

Rawski, M., J. Mazurkiewicz, B. Kierończyk, and D. Józefiak. 2021. "Black soldier fly full-fat larvae meal is more profitable than fish meal and fish oil in siberian sturgeon farming: The effects on aquaculture sustainability, economy and fish GIT development." *Animals* 11 (3). doi:10.3390/ani11030604.

Ray, B. 1992. "Cells of lactic acid bacteria as food biopreservatives." In: B. *Ray and M. A. Daeschel (Eds.), Food Biopreservatives of Microbial Origin.* CRC Press.

Ray, B., and W. E. Sandine. 1992. "Acetic, propionic and lactic acids of starter culture bacteria as biopreservatives." In: B. *Ray and M. A. Daeschel (Eds.), Food Biopreservatives of Microbial Origin.* CRC Press.

Robertson, R. P. 2023. "Nrf2 and antioxidant response in animal models of type 2 diabetes." *Int J Mol Sci* 24 (4). doi:10.3390/ijms24043082.

Ruckli, A. K., S. Hörtenhuber, S. Dippel, P. Ferrari, M. Gebska, M. Heinonen, J. Helmerichs, C. Hubbard, H. Spoolder, A. Valros, C. Winckler, and C. Leeb. 2024. "Access to bedding and outdoor runs for growing-finishing pigs: Is it possible to improve welfare without increasing environmental impacts?" *Animal* 18 (5):101155. doi:10.1016/j.animal.2024.101155.

Sabioni, R. E., E. K. Lorenz, J. E. P. Cyrino, and H. Volkoff. 2022. "Feed intake and gene expression of appetite-regulating hormones in Salminus brasiliensis fed diets containing soy protein concentrate." *Comp Biochem Physiol A Mol Integr Physiol* 268:111208. doi:10.1016/j.cbpa.2022.111208.

Saengrung, J., A. Bunnoy, X. Du, L. Huang, R. An, X. Liang, and P. Srisapoome. 2023. "Effects of ribonucleotide supplementation in modulating the growth of probiotic *Bacillus subtilis* and the synergistic benefits for improving the health performance of Asian seabass (*Lates calcarifer*)." *Fish Shellfish Immunol* 140:108983. doi:10.1016/j.fsi.2023.108983.

Samat, N. A., F. M. Yusoff, N. W. Rasdi, and M. Karim. 2020. "Enhancement of live food nutritional status with essential nutrients for improving aquatic animal health: A review." *Animals (Basel)* 10 (12). doi:10.3390/ani10122457.

Santos, A., F. J. Giráldez, C. Valdés, E. Trevisi, L. Lucini, J. Frutos, and S. Andrés. 2018. "Milk replacer restriction during early life impairs the live body weight and progesterone patterns of ewe lambs during the replacement period." *J Dairy Sci* 101 (9):8021–31. doi:10.3168/jds.2018-14648.

Satoh, H., R. Fukumori, R. Kumano, M. Kamata, K. Shimada, K. Izumi, and S. Oikawa. 2024. "Effects of starch content of calf starter on rumen properties and blood concentrations of metabolites and hormones in dairy calves under a high plane of milk feeding." *Anim Sci J* 95 (1):e13927. doi:10.1111/asj.13927.

Scholten, M. C. Th, de Boer I. J. M., Gremmen B., and C. and Lokhorst. 2013. "Livestock farming with care: Towards sustainable production of animal-source food." *NJAS: Wageningen J Life Sci* 66 (1):3–5. doi:10.1016/j.njas.2013.05.009.

Schrobback, P., C. Gonzalez Fischer, D. Mayberry, and M. Herrero. 2023. "On-farm investments into dairy cow health: Evidence from 15 case study countries." *Front Vet Sci* 10:1288199. doi:10.3389/fvets.2023.1288199.

Scott, K., D. F. Kelton, T. F. Duffield, and D. L. Renaud. 2019. "Risk factors identified on arrival associated with morbidity and mortality at a grain-fed veal facility: A prospective, single-cohort study." *J Dairy Sci* 102 (10):9224–35. doi:10.3168/jds.2019-16829.

Seelam, J. S., M. Fernandes de Souza, P. Chaerle, B. Willems, E. Michels, W. Vyverman, and E. Meers. 2022. "Maximizing nutrient recycling from digestate for production of protein-rich microalgae for animal feed application." *Chemosphere* 290:133180. doi:10.1016/j.chemosphere.2021.133180.

Shurson, G. C. 2025. "We can optimize protein nutrition and reduce nitrogen waste in global pig and food production systems by adopting circular, sustainable, and one health practices." *J Nutr* 155 (2):367–77. doi:10.1016/j.tjnut.2024.12.014.

Shurson, G. C., and P. E. Urriola. 2022. "Sustainable swine feeding programs require the convergence of multiple dimensions of circular agriculture and food systems with one health." *Anim Front* 12 (6):30–40. doi:10.1093/af/vfac077.

Shurson, G. C., P. E. Urriola, and Y. T. Hung. 2022. "Too much of a good thing: Rethinking feed formulation and feeding practices for zinc in swine diets to achieve one health and environmental sustainability." *Animals* 12 (23). doi:10.3390/ani12233374.

Shurson, G. C., P. E. Urriola, and J. L. G. van de Ligt. 2022. "Can we effectively manage parasites, prions, and pathogens in the global feed industry to achieve one health?" *Transbound Emerg Dis* 69 (1):4–30. doi:10.1111/tbed.14205.

Siddiqui, S. A., D. N. Adli, W. S. Nugraha, B. Yudhistira, F. V. Lavrentev, S. Shityakov, X. Feng, A. Nagdalian, and S. A. Ibrahim. 2024. "Social, ethical, environmental, economic and technological aspects of rabbit meat production: A critical review." *Heliyon* 10 (8):e29635. doi:10.1016/j.heliyon.2024.e29635.

Siegford, J. M. 2013. "Multidisciplinary approaches and assessment techniques to better understand and enhance zoo nonhuman animal welfare." *J Appl Animal Welfare Sci* 16 (4):300–18. doi:10.1080/10888705.2013.827914.

Sørensen, S. L., Y. Park, Y. Gong, G. K. Vasanth, D. Dahle, K. Korsnes, T. H. Phuong, V. Kiron, S. Øyen, K. Pittman, and M. Sørensen. 2020. "Nutrient Digestibility, growth, mucosal barrier status, and activity of leucocytes from head kidney of atlantic salmon fed marine- or plant-derived protein and lipid sources." *Front Immunol* 11:623726. doi:10.3389/fimmu.2020.623726.

Spinler, M. S., J. C. Woodworth, M. D. Tokach, R. D. Goodband, J. M. DeRouchey, H. L. Frobose, A. Uitermarkt, and J. T. Gebhardt. 2024. "Evaluation of precision feeding standardized ileal digestible lysine and other amino acids to determine and meet the lactating sow's requirement estimates." *J Anim Sci* 102. doi:10.1093/jas/skae285.

Stenberg, O. K., E. Holen, L. Piemontese, N. S. Liland, E. J. Lock, M. Espe, and I. Belghit. 2019. "Effect of dietary replacement of fish meal with insect meal on *in vitro* bacterial and viral induced gene response in Atlantic salmon (*Salmo salar*) head kidney leukocytes." *Fish Shellfish Immunol* 91:223–32. doi:10.1016/j.fsi.2019.05.042.

Stevenson, M. A., R. L. Sanson, A. O. Miranda, K. A. Lawrence, and R. S. Morris. 2007. "Decision support systems for monitoring and maintaining health in food animal populations." *New Zealand Veter J* 55 (6):264–72. doi:10.1080/00480169.2007.36780.

Sturino, J. M., M. Rajendran, and E. Altermann. 2013. "Draft genome sequence of the pediocin-encoding biopreservative and biocontrol strain pediococcus acidilactici D3." *Genome Announc* 1 (3). doi:10.1128/genomeA.00208-13.

Takayasu, S., T. Sakurai, S. Iwasaki, H. Teranishi, A. Yamanaka, S. C. Williams, H. Iguchi, Y. I. Kawasawa, Y. Ikeda, I. Sakakibara, K. Ohno, R. X. Ioka, S. Murakami, N. Dohmae, J. Xie, T. Suda, T. Motoike, T. Ohuchi, M. Yanagisawa, and J. Sakai. 2006. "A neuropeptide ligand of the G protein-coupled receptor GPR103 regulates feeding, behavioral arousal, and blood pressure in mice." *Proc Natl Acad Sci USA* 103 (19):7438–43. doi:10.1073/pnas.0602371103.

Thakur, J. S., A. Rana, R. Kaur, R. Paika, S. Konreddy, and M. Wiktorowicz. 2025. "Situational analysis of human and agricultural health practice: One Health and antibiotic use in an indigenous village in rural Punjab, India." *One Health* 20:100946. doi:10.1016/j.onehlt.2024.100946.

Tizard, M., E. Hallerman, S. Fahrenkrug, M. Newell-McGloughlin, J. Gibson, F. de Loos, S. Wagner, G. Laible, J. Y. Han, M. D'Occhio, L. Kelly, J. Lowenthal, K. Gobius, P. Silva, C. Cooper, and T. Doran. 2016. "Strategies to enable the adoption of animal biotechnology to sustainably improve global food safety and security." *Transgenic Res* 25 (5):575–95. doi:10.1007/s11248-016-9965-1.

Tuma, A. 2024. "Mastering animals hygiene: Essential procedures for animal welfare by Robin Adamas." *Allied Sci J Vet Med* 8 (1):184.

van Houtert, M. F., and A. R. Sykes. 1999. "Enhancing the profitability of pasture-based dairy production in the humid tropics through improved nutrition." *Prev Vet Med* 38 (2–3):147–57. doi:10.1016/s0167-5877(98)00120-2.

van Wagenberg, C. P. A., Y. de Haas, H. Hogeveen, M. M. van Krimpen, M. P. M. Meuwissen, C. E. van Middelaar, and T. B. Rodenburg. 2017. "Animal board invited review: Comparing conventional and organic livestock production systems on different aspects of sustainability." *Animal* 11 (10):1839–51. doi:10.1017/s175173111700115x.

Varela-Ortega, C., I. Blanco-Gutiérrez, R. Manners, and A. Detzel. 2022. "Life cycle assessment of animal-based foods and plant-based protein-rich alternatives: A socio-economic perspective." *J Sci Food Agric* 102 (12):5111–20. doi:10.1002/jsfa.11655.

Verga, M., and M. Michelazzi. 2009. "Companion animal welfare and possible implications on the human–pet relationship." *Italian J Anim Sci* 8 (sup1):231–40. doi:10.4081/ijas.2009.s1.231.

Vyas, D., C. D. Nelson, J. J. Bromfield, P. Liyanamana, M. Krause, and G. E. Dahl. 2020. "MILK symposium review: identifying constraints, opportunities, and best practices for improving milk production in market-oriented dairy farms in Sri Lanka." *J Dairy Sci* 103 (11):9774–90. doi:10.3168/jds.2020-18305.

Wang, X., H. Wang, Y. Liu, G. Li, Y. Yang, C. Wang, S. Gong, D. He, S. Chen, and H. Jia. 2024. "Optimizing low-protein diets with edible dock powder: Integrated effects on growth performance, slaughter quality, organ weights, muscle quality, and cecal microbiota in growing Sanhua geese." *Poult Sci* 103 (12):104476. doi:10.1016/j.psj.2024.104476.

Wang, Y., H. Hou, H. Luo, J. Xun, C. Ma, H. Yang, D. Bai, S. Yousuf, H. Lyu, T. Zhang, X. Wan, X. Yao, T. Ma, Y. Zhou, Z. Zhu, M. Zeng, S. An, Q. Bai, Y. Bai, G. Cao, T. Cao, Y. Cao, C. Chang, L. Chang, B. Chen, D. Chen, D. Chen, H. Chen, J. Chen, J. Chen, W. H. Chen, X. Chen, Y. Chen, Z. Chen, C. Cheng, Q. Cheng, X. J. Dai, C. Deng, F. Deng, J. Deng, C. S. Dong, L. Dong, L. Duan, Y. Duan, Q. Fan, C. Fang, T. Fang, W. Fang, Z. Fang, M. Fu, M. Fu, C. Gao, H. Gao, W. Gao, X. Gao, Y. Z. Gao, Y. Geng, W. Gong, S. Gu, X. Gu, Z. Gu, J. W. Guo, J. Guo, Q. Guo, X. Guo, X. Guo, D. Han, Z. Han, Y. Hao, J. He, J. He, J. He, R. He, G. Hou, B. Hu, H. Hu, Y. Hu, Y. Hu, Y. Hu, G. Huang, H. Huang, J. Huang, S. Huang, B. Jia, X. Jian, C. Jiang, K. Jiang, L. Jiang, S. Jiang, J. Y. Jiao, H. Jin, J. Jin, S. Kong, X. Lai, Y. Leng, B. Li, B. Li, F. Li, H. Li, H. Li, J. Li, K. Li, L. Li, L. Li, M. Li, P. Li, W. Li, W. Li, X. Li, L. Xuemeng, Y. Li, Y. Li, Z. Li, L. Liang, R. Liang, Z. Liang, Q. Liu, D. Liu, H. Liu, J. Liu, L. Liu, L. Liu, M. Liu, R. Liu, S. Liu, T. Liu, W. Liu, W. Liu, X. Liu, Y. Liu, Y. Liu, Y. Liu, Y. Liu, Z. Liu, Z. Liu, Z. Liu, C. Long, Y. Long, C. Lu, C. Lu, C. Lu, Q. Lu, Y. Luan, P. Luo, S. Luo, N. Ma, X. Y. Ma, Y. Ma, W. Mao, Y. Meng, Y. Ni, Y. Ni, K. Ning, D. Niu, K. Peng, Z. Peng, X. Qian, Z. Qiu, H. Qu, Z. Qu, Y. Ren, Z. Ren, Y. Shen, L. Shi, L. Shi, W. Shi, Y. Shi, T. Song, X. Song, X. Song, X. Song, Q. Su, Y. Su, L. Sun, Q. Sun, T. Sun, Y. Sun, H. Tang, W. Tang, T. Yu, S. Tian, S. Wang, B. Wang, C. Wang, W. Jin, L. Wang, L. Wang, L. Wang, M. Wang, M. K. Wang, P. Wang, S. Wang, S. Wang, X. Wang, X. Wang, M. Wei, Y. Wei, Y. Wei, Y. Wei, C. Wen, X. Wen, L. Wu, S. Wu, Y. Wu, S. Xia, X. Xia, Y. Xia, X. Xiang, C. Xiao, W. Xiao, Y. Xiao, R. Xie, R. Xing, H. Xu, W. Xu, Z. Xu, H. Xue, C. Yan, Q. L. Yan, S. Yan, X. Yan, M. Yang, Y. Yang, Z. Yang, Z. Yang, G. Yao, Y. Yao, X. Yi, C. Yin, M. Yin, S. Yu, Y. Yu, Y. Yu, F. Yuan, S. L. Zhai, B. Zhang, C. Zhang, F. Zhang, F. L. Zhang, H. Zhang, J. Zhang, J. Zhang, K. Zhang, L. Zhang, L. Zhang, L. Zhang, M. Zhang, Q. Zhang, R. Zhang, T. Zhang, T. Zhang, W. Zhang, Y. Zhang, Y. Zhang, Y. Zhang, Z. Zhang, Z. Zhang, Z. F. Zhang, B. Zhao, Y. Zhao, Y. Zhao, Z. Zhao, D. Zheng, Y. Zheng, W. Zhi, J. Zhong, X. Zhong, W. Zhou, X. Zhou, Z. Zhou, Z. Zhou, C. Zhu, F. Zhu, X. Zhu, Y. Zou, H. Zhou, L. Lei, Y. Bi, H. Shi, H. Z. Sun, S. Jin, W. Ren, L. Dai, X. Wang, C. Lan, H. Liu, S. J. Liu, Y. Yin, C. L. Shi, R. Y. Gan, F. Zhao, J. Yu, T. Chen, X. Hong, H. Yang, B. Zhang, S. Chen, X. Li, Y. Gao, and Y. X. Liu. 2024. "iMeta Conference 2024: Building an innovative scientific research ecosystem for microbiome and one health." *Imeta* 3 (6):e251. doi:10.1002/imt2.251.

Wang, Y., H. Quan, X. Li, Q. Li, M. A. Haque, Q. Shi, Q. Fu, and C. He. 2021. "Contamination with fumonisin B and deoxynivalenol is a threat to egg safety and contributes to gizzard ulcerations of newborn chickens." *Front Microbiol* 12:676671. doi:10.3389/fmicb.2021.676671.

Warner, D., G. M. Dallago, O. W. Dovoedo, R. Lacroix, H. A. Delgado, R. I. Cue, K. M. Wade, J. Dubuc, D. Pellerin, and E. Vasseur. 2022. "Keeping profitable cows in the herd: A lifetime cost-benefit assessment to support culling decisions." *Animal* 16 (10):100628. doi:10.1016/j.animal.2022.100628.

Wertz, A. E., T. J. Knight, A. Trenkle, R. Sonon, R. L. Horst, E. J. Huff-Lonergan, and D. C. Beitz. 2004. "Feeding 25-hydroxyvitamin D3 to improve beef tenderness." *J Anim Sci* 82 (5):1410–8. doi:10.2527/2004.8251410x.

Will, K. J., E. S. Magalhaes, C. A. A. Moura, G. Trevisan, G. S. Silva, A. P. G. Mellagi, R. R. Ulguim, F. P. Bortolozzo, and D. C. L. Linhares. 2024. "Risk factors associated with piglet pre-weaning mortality in a Midwestern U.S. swine production system from 2020 to 2022." *Prev Vet Med* 232:106316. doi:10.1016/j.prevetmed.2024.106316.

Winkelman, L. A., T. H. Elsasser, and C. K. Reynolds. 2008. "Limit-feeding a high-energy diet to meet energy requirements in the dry period alters plasma metabolite concentrations but does not affect intake or milk production in early lactation." *J Dairy Sci* 91 (3):1067–79. doi:10.3168/jds.2007-0434.

Yang, L., X. He, D. Zhi, Y. Xue, X. Gong, K. Dong, and Y. Tian. 2024. "Melittin promotes dexamethasone in the treatment of adjuvant rheumatoid arthritis in rats." *Front Pharmacol* 15:1338432. doi:10.3389/fphar.2024.1338432.

Yeh, Y. L., P. C. Chou, Y. H. Chen, L. S. Lai, T. K. Chung, R. L. Walzem, S. Y. Huang, and S. E. Chen. 2020. "Dietary supplementation of 25-hydroxycholecalciferol improves cardiac function and livability in broiler breeder hens-amelioration of blood pressure and vascular remodeling." *Poult Sci* 99 (7):3363–73. doi:10.1016/j.psj.2020.03.015.

Yu, H., K. Lee, and G. Morota. 2021. "Forecasting dynamic body weight of nonrestrained pigs from images using an RGB-D sensor camera." *Transl Anim Sci* 5 (1):txab006. doi:10.1093/tas/txab006.

Zarantoniello, M., B. Randazzo, G. Gioacchini, C. Truzzi, E. Giorgini, P. Riolo, G. Gioia, C. Bertolucci, A. Osimani, G. Cardinaletti, T. Lucon-Xiccato, V. Milanović, A. Annibaldi, F. Tulli, V. Notarstefano, S. Ruschioni, F. Clementi, and I. Olivotto. 2020. "Zebrafish (*Danio rerio*) physiological and behavioural responses to insect-based diets: A multidisciplinary approach." *Sci Rep* 10 (1):10648. doi:10.1038/s41598-020-67740-w.

Zargari, A., M. Mazandarani, R. Safari, H. Hoseinifar, and A. Hedayati. 2024. "Modulation of toxic effects of ammonia on growth, pathology of liver and kidney tissues and relative expression of GH and IGF-1 genes by CoQ(10) supplementation in *Oncorhynchus mykiss*." *Fish Physiol Biochem* 50 (4):1711–29. doi:10.1007/s10695-024-01365-y.

Zhang, L., Y. Bai, J. Tao, S. Yang, C. Tu, L. Liu, X. Huang, L. Li, and Z. Qin. 2024. "Effects of feeding chicken egg yolk antibodies on intestinal cell apoptosis, oxidative stress and microbial flora of tilapia (*Oreochromis niloticus*) infected with *Streptococcus agalactiae*." *Fish Shellfish Immunol* 150:109596. doi:10.1016/j.fsi.2024.109596.

Zhao, Y., Y. Hou, F. Liu, A. Liu, L. Jing, C. Zhao, Y. Luan, Y. Miao, S. Zhao, and X. Li. 2016. "Transcriptome analysis reveals that vitamin a metabolism in the liver affects feed efficiency in pigs." G3 6 (11):3615–24. doi:10.1534/g3.116.032839.

Zong, B., Y. Zhu, M. Liu, X. Wang, H. Chen, Y. Zhang, and C. Tan. 2022. "Characteristics of mycoplasma hyopneumoniae strain es-2 isolated from Chinese native black pig lungs." *Front Vet Sci* 9:883416. doi:10.3389/fvets.2022.883416.

Żuk-Gołaszewska, K., R. Gałęcki, K. Obremski, S. Smetana, S. Figiel, and J. Gołaszewski. 2022. "Edible insect farming in the context of the EU regulations and marketing: An overview." *Insects* 13 (5). doi:10.3390/insects13050446.

10 Inevitability of Balanced Life and the Growth of Farming Animals

Amro Abd Al Fattah Amara,
Mounir Mohamed Salem-Bekhit,
Khairy Mohamed Abdallah Zoheir,
Nawal Abd El-Baky,
Mohamed Morsi Mohamed Ahmed,
and Arti Gupta

ABBREVIATIONS

ARG	Antibiotic resistance genes
BSE	Bovine spongiform encephalopathy
CFFS	Crowded fast farming style
TSFS	Traditional slow farming style

10.1 INTRODUCTION

The human understands that the natural balance is not new. Yes, nowadays, humans have learned to preserve their foods by applying different simple techniques such as smoking (Liu et al. 2018; Dias et al. 2020; Lingbeck et al. 2014), salting (Lück 1985; Sapeika 1973), drying (Lipan et al. 2020; Murphy et al. 2017; Singh and Shalini 2016), fermenting (Bourdichon 2012), keeping seeds in their covers, and adding some herbaceous plants (e.g., black peppermint) to preserve food. Meanwhile, a significant change has happened in the past decade. Nowadays, many tools are used in food preservation (Kumar et al. 2020; Margalho et al. 2021; Sridhar et al. 2021). The best among them are those that use safe, natural products that are compatible with our bodies (Margalho et al. 2021; Ray 1992; Ray and Sandine 1992; Sturino, Rajendran, and Altermann 2013).

In the traditional Egyptian farming style, some practices show humans putting the seeds by force in the birds' mouths (e.g., ducks, ooze, and the like) and forcing them to engulf the seeds. But that practice was followed only a few weeks before their slaughtering to give more weight. In other practices, some farmers mix salt in the animal's food (non-accepted practices by the society) to force the animals to drink more water, aiming to gain more weight. But in the traditional farming style,

DOI: 10.1201/9781003452805-10

there are no use of hormones (Arieli et al. 2001), no use for long day-light time (Nosratimovafagh, Fereidouni, and Krujatz 2022) to keep animals feed, no crowding, etc. In fact, most animals are essential parts of the farmers' activities, so they give them enough food, do not exhaust them by extra work, and keep their growing place dry, clean, well aerated, and healthy.

One could say that some of the big animals are a source of enjoyment for the children. For example, they take the buffalo to the nearest water source suitable for swimming and swim beside it. The same happens with the horses. The most enjoyable part for children when they have access to the countryside is riding a donkey. They ask their parents when they will go to the countryside. They are interested in riding the donkey! Additionally, there are many traditional and religious stories about a human or a group of people who engage animals in a particular story, or an imaginative roman is built to tell stories about the animals, for example, the famous philosophic book *Kalila wa-Demna*. It is a great philosophic Indian book (translated and improved by the Araba) with short philosophic stories written to teach an Indian king how to choose the correct solutions without stimulating his anger.

In other parts of the world, in the Middle East, the slaughtering of animals is a practice that needs some sort of expertise. There is a restricted and precise instruction about how to slaughter the animals. Some believe and practice that, but only after getting approval from an old person to keep this practice in the hands of a narrow number of persons. An order is given to make the knife so sharp; give the animal water before the slaughtering process; avoid the animal from seeing the knife by any means; and provide a lot of details. The slaughtering must be at certain points in the neck to keep the animal breathing (till death). After slaughtering, the animal must be left free (without captivity) immediately to die in the most humane, relaxed condition. Many today might not accept the slaughtering process at all, or at least do not slaughter by themselves, but at least the process has been optimized to give the most available acceptable conditions.

Some give each animal a name. Other farming animals are grown in a free, open area where they are only guided by sound or by a piece of wood. Collectively, the farmers in the old world before the Industrial Revolution were satisfied with their regular, happy life. A few have been trying to change this system for hundreds of years (Lange et al. 2020; Truong et al. 2021). Yes, there are some small running businesses, but they are totally not comparable with those run by the mega companies nowadays. Buffalos or cows work with the family on the farm when they pass the period where they are powerful and before they become really old, and they are stopped from working on the farm and are given more foods to build their body and to give weight (to be sealed and slaughtered).

It is some sort of drama when one animal disappears from the farm. In many cases, children are given some incorrect explanations to avoid any bad memories. Some believe that if you harm an animal, you must be reclaimed on the day of reclamation. The respect for farm animals is very clear in most of the old world. In countries like India, the respect is exponential. So, what happened to the animal welfare (Green and Mellor 2011; Millman 2009; Rogers-Brambell 1965) and why such sudden big changes and deviation!

10.2 THE NATURE IS BALANCED BY DEFAULT

10.2.1 Enough Genes

Things and beings are diverse. In nature, this diversity is in certain equilibrium and harmony that keep all their components in certain balance. Perhaps that is because everything, either materialistic or growable, is perfectly created to react in a relaxed way with the other components. For example, antagonistic microbes could be isolated from the same sample size. The antibiotic resistance genes, which were thought early to be due to evolution, are proven to be part of the microbes that produce the antibiotics to enable them to survive against their own product. Yes, changes happened everywhere, but they were either reversible or irreversible. Such changes could affect our genes (Amara and Shibl 2015; Amara, Shibl, and Tawfik 2012), that is, by the force of selection due to certain conditions, but the original genes usually still exist and are ready to appear again at the correct time. Or that everything (every gene) existsbut each one appears in certain cases or conditions. Seldom, a certain creature is eradicated without a force. For example, some individuals survived after a very fatal disease like smallpox, even if it reaches non-immunized isolated groups. In other places, many individuals survived, and the first treatment is taken from the skin granules (due to the response of the immunized person) and is used to immunize the children. Variolation is an old practice, where the weakened virus (by heat or by the immune system response) is given to the nonspecific part of the hard hand muscle. Epidemic diseases have appeared on the earth hundreds of times, but have not eradicated humans. The creatures have enough power to survive. Microbes, except viruses, can live, reproduce, die, and evacuate naturally. From nature and its activities, one could learn a lot (Amara et al. 2014; Sheweita et al. 2019; Amara, Salem-Bekhit, and Alanazi 2013).

10.2.2 The Power of Nature

Nature has additional power to correct any damage (for certain limits) globally. However, changes that are made by humans in one or more than one of the nature's components might lead to contradiction. That is because some did not understand the roles that govern the nature and its components. Only three roles are highlighted as follows. One major role that could be addressed here is that each creature has an enemy (or anomaly). Each creature has its specific enemy that keeps its number at a constant level. If you change or eradicate this enemy, nature will become your enemy. Another role in nature is to survive and keep other creatures survive as well, without exhausting nature. If one applies those two roles, real-world examples can be given. Carnivorous animals hunt only when they need to eat. A group of hunters will focus on one animal to satisfy their demand for certain days. Who disturbs this harmony in most cases? In fact, the human who often intervenes with a superior look, unaware of the consequences of matters , reaps the consequences of his wrong actions, which do not affect him alone but rather affect everyone! Carnivores are responsible for limiting the numbers of herbivores. A job allows plants to grow appropriately, a process that sustains the lives of the herbivores themselves. When carnivorous animals such as the wolves were eliminated from a balanced natural area, herbivores multiplied in a way that resulted in eliminating many herbaceous plants, which in reverse turn led to eliminating their eater, the herbivores themselves, and caused drought. Finally, the equilibrium takes place. If you eradicate the hunter, the carnivores will eradicate the herbs and eradicate themselves and the place will be totally corrupt. Because there is no balance.

In a few cases, animals are reported to make changes in the biodiversity. But after deep inspection, one will mostly discover that those animals have been foreign to these habitats or they are domestic ones that come back to the wild nature but without the memory of their similar relatives (wild type trip). For examples, when domestic dogs are brought to Mauritius, an island (Madagascar) of the Indian Ocean, they eradicate the DODO (*Raphus cucullatus*), because they have eaten their eggs. The third role is that the equilibrium is not based on a single role. For example, in wildlife, there is not one type of herbivore animal or one type of carnivore as well. However, one could remark that herbivores are compatible with each other, while carnivores try to coexist far from those who are stronger than them, either from the same species or from different species. You find them all living in groups of different numbers, each group's members complement each other and defend each other. These beings flowed in a wonderful system based on delicate balance, so that if left without human intervention, they could coexist in nearly constant numbers. This diversity and harmony could be applied not only to animals but also to plants, insects, and all living creatures. For example, the use of antibiotics leads to the accumulation of antibiotic residues in nature and creates antibiotic resistance. Only antibiotic-resistant microbes will survive first and then they become dominant. For more details about the non-balanced conditions that occurred due to the excessive use of the antibiotic, refer to Amara et al. (2024) plus the related text given below.

During their life on earth, humans domesticate some wild animals, such as the dogs to protect themselves and the cats to kill rats and snakes. Also, some animals became domesticated for food, such as buffalo, cows, sheep, and birds, and for riding, such as horses and donkeys. These animals, which were probably brought from the wild, accompanied them on their travels and were a suitable source of food, clothing, travel, covers, and for making tents and woolen and felt textiles. This balanced, natural, monotonous life continued like this for thousands of years and did not deviate from its natural scope except for limited times.

10.2.3 The Significant Changes in the Natural Balanced Life after the Industrial Revolution

The Industrial Revolution and the First and Second World Wars represented a radical change in human life generally and in animal welfare specifically. Millions of workers, soldiers, travelers are in demand for food due to their special day conditions. As a result, many new practices are invented. Millions of employers are eating foods that are not prepared in-house. The wife who is returning to her home tired and exhausted needs a fast solution to prepare the food. To satisfy all those who either in special conditions or do not have enough time, the crowded farming style is invented. But that was also not enough in the first place. A research was conducted to optimize every single step included in the crowded farming process and increase the animal output of the targeted product such as the animal food components, growing style, the overall processes, prophylaxes, treatment, and vaccination.

This is a nice explanation for the changes that have occurred or the nice face of the coin. The other face concerns with the profit that interferes with many of our modern practices, like the huge production of antibiotics, the control of landscape that used to produce human food, degrading the family farm style, and controlling the market, which are not in the scope and the aim of this chapter.

10.3 OUR FOCUS ON THE BRANCHES AND IGNORANCE TO THE ROOT OF THE TREE (THE PROBLEM)

10.3.1 Example No 1: The Fast-Food

Many blamed the fast-food producers and charge them with the responsibility for our health problems. But they never force us to buy or to eat their products. In fact, humans have dropped many activities that cut the way against any support but keep only one support represented in the fast-food producer (face to face). No mother, no grandmother, and in many cases no wife or, in another word, no one responsible for the demands of a family in its knownsense. In Egypt, these new conditions due to excessive modernization are called "the isolated island" or we live in isolated islands even if we live in the same home. If such cases have happened to you, to our children, and to our family members, what could one expect for the animals or the family farm animals.

10.3.2 Example No 2: Modernization or the Kick-Off of the Farm Animals!

One great mistake in the countryside that has happened intentionally or non-intentionally is substituting the semi-manual irrigation style, where animals are used in the irrigation system with machines. In fact, there are two major side effects; the first mistake is the huge amount of petroleum products used to run the irrigation machines/year. The second mistake is that the farm animals are getting out of the equation and over time many farmers are getting rid of them. The old style forces the farmer to select the best animal able to help in irrigating the land. It is the most unplanned selections that happened in nearly every place. Also, what could happen in the case of the absence of the petroleum products! This is another example of how humans could interfere unwisely with things that have been installed for hundreds of years and prove safety and success. If one puts the consequences in order, in fact, the land that is prepared to feed humans is used first to produce products to satisfy the demand of the crowded farming style. Apparently, that was not enough! Perhaps, somewhere a decision is taken to kick away the family farm animal! No one could answer, prove or disprove.

10.3.3 Example No 3: The Hidden Role of Antibiotics

In other words, the cities became the place for factories and luxurious life, and people abandoned the countryside, the grazing, and the agricultural areas that used to provide everyone with food. The new industrial model based on giant factories that eliminated small manufacturers finally distracted the small producers. The earlier production model was concerned with a specific number of animals and birds which did not take a lot from the producer's time to allow them to live healthy in a healthy environment. Even in case of losses, one could recover such losses in a short time! In contrast, giant farms were created to produce thousands of birds and thousands of tons of meat. Problems from this intensive production began to appear in terms of the incidence of diseases and their transmission to humans and the surrounding environment. Luckily, a hero appeared. It is the antibiotics, which saved this industry from self-destruction. But man was not satisfied with this success, nor was he satisfied with the positive factors and privileges provided to him by nature (the antibiotic) to correct his mistakes. He almost kicked away natural antibiotics and produced their synthetic counterparts, which were produced in very large quantities. The use of antibiotics has extended to many fields in agriculture, animal production, and, of course, for human treatment. All of this led to contaminating nature with unusual amounts of antibiotics. This led to disrupting the natural balance of microbes, as we find those who have genes that are resistant to antibiotics or have developed forms of their cellular structure to prevent antibiotics from reaching their target or being expelled quickly after the entry become dominant. One could add the biofilm producer, which reduced the microbes' sensitivity to antibiotics, and the ability to consume food and other compounds. Again, one change does a cascade of changes, either specific or nonspecific. The above-described issues highlight the importance of the inevitability of balanced life (Amara 2013).

10.4 OUR FOOD THE MOST IMPORTANT MEDICAL ISSUE AND THE DE *NEVO* TOXIN

The food is a critical part of our life. No life without food. Humans can survive for a countable days without food. They are more in number than one can survive without water! Many people describe that our food is the main source of illness, even though the food in nature is healthy and without any kind of the newly elevated toxins, or, in other words, the toxin *de nevo* of our modern civilization. Yes, even the food is completely healthy, it could be harmful if consumed in quantities that exceed our body's ability and capacity to manage its digestion, catabolism, and anabolism. So, what about food that is already toxic in its nature? Or that are contaminated during the many steps during its processing like storage, cultivation, harvesting, transportation, restorage, industrial processing, and preservation.

Why new medicines are invented every day to treat illnesses caused by the newly produced toxins and resistant microbes. Both types of products that cause illness or side effects and those that are proposed to cure one or more than one of the effects that happened due to such side effects collectively are accumulated in nature and finally find their way to our bodies.

10.4.1 Some Survivals

In fact, some civilizations that are vegetarian in their attitude are lucky and survive by being far away from nearly 50% of the available toxins in the foods. If one is a vegetarian, there is no need to force animals to grow faster using antibiotics, drugs, and even medications to treat the expected side effects. In fact, if one is a vegetarian, there is no need even for reading about the addressed issues in this chapter. Shouldn't all become vegetarians even those who have no restrictions to eat living creature meat and products could practice that to force the correction to happen. Exactly, we are all forcing incorrect style to satisfy our unsatisfied demands. In fact, there will be no objection from any religions to turn us all to be vegetarian for a time because the priority in any religions is to save the human (us), the animals, the nature, etc.

Other groups who are partially saved are those who sloughed animals correctly. They have escaped from other respective parts of the toxins. Some other groups who are either so primitive or isolated from our modern civilization are still utilizing natural, healthy foods. Perhaps in future, if humans succeed in solving the different problems due to food contamination with different chemical compounds or antibiotics, a question will be asked. How such generation able to eat food contaminated with toxic residues as their documents prove? It must be a typing mistake! Or, if more toxic compound enters our foods they will say; "their food was not containing enough toxins!" In fact, our modern civilization could not be totally or completely reclaimed, but some of us who are forcing all the available power in their hand to gain more and more profit, control, monopoly, etc., with a complete ignorance of the long-run side effects are responsible for that. As a result, everyone is invited to protect themselves, their family members, and the like, as well as to protect their animals and others who might have been harmed during some unwise practices, like using harmful chemical compounds during a cultivation process without sustainable control, unwise animal farming, non-sustainable microbial control or infection treatments, uncontrolled gene transfers, and the like. One should differentiate between the amount of healthy food he consumed and the different levels of harmful residues that might contaminate his food. In some cases, small quantities of residues that could not be detected by any means are more harmful if food is consumed for long periods. In such a case, the damage will probably be irreversible.

Animals, in many cases, are used instead of humans to prove or disprove the food quality and safety. Not only do scientists use animals as an alternative to humans to avoid direct toxicity but also some animal producers feed animals with foods that might be unacceptable based on their quality or that are human food products that were already expired or, in the best scenario, on their way to expire. Many might consider that this practice is safe and beneficial and omits the risk of being utilized by humans but in fact, in many cases, this argument is incorrect, while the cycle of the toxins will finally reach our plates. In one or another way, humans will indirectly be the victim.

10.4.2 An Invitation to React More Natural

The concept of the inevitability of balanced life might be an acceptable solution to protect the development that happened in our life and to guarantee healthy food (Amara 2013). It is also a pre-invitation to sense what is compatible with the nature generally and the nature of farming animals specifically, and what is not compatible! Farm animals like different kinds of poultry, cows, buffalo, sheeps, goats, etc., are mostly herbivores. So, why we supply them with food that contains animal products. It is an invitation to react more naturally even when things look like they are non-harmful but are deviated from the natural behavior of animals and humans as well. Perhaps, and at least for humans, many of the forbidden attitudes are due to the negative experiences of a past generation where the advices are transferred to us through time. For that, our memory could sense, for some extent, what is correct and what is incorrect. Did anyone sense that giving a cow an animal product is the correct choice!

If food from a different nature could harm herbaceous animals, what about food that is already contaminated by different undesired residues. Now one question might be addressed: when we decided to accept eating food and feed our animals as well with food that contains toxins like synthetic antimicrobial agents, drugs, residues from the fertilization process (chemical fertilization), or non-natural chemical compounds? And could we stop (now)?

10.4.3 We Repeated the Same Mistake(s)

The mistake made by Adam and Eve and their excessive first generation, when they ate from the forbidden tree, was a harsh lesson that was passed down through many generations and was reflected in many practices that show man's attempt to control himself and his wishes. This is clearly evident in their first place of residence in the old world and then represented in different forms in the rest of the neighboring regions until it completely disappears in the regions that are geographically and culturally distant. To control himself (human), many practices were practiced, the most obvious and perhaps the most correct one is practicing fasting, which does not take one form but rather varies. Some fast from a specific type of food, others fast for a specific time from some types of food, and others fast for a specific time from all types of food, etc.

There is no doubt that these attempts that strengthen the ability to control our hopes and demands will be reflected in one way or another in the ability to make the right decision. If one understands the concept that a large amount of food will harm his/her body that will enable them to pass such understanding to the animals. Animals, from a scientific point of view, are a complete biological system similar to humans in most aspects. In another angle, if human utilizes only his bodily necessity to be powerful and healthy, that, of course, will be reflected in the amount of food produced and consumed, and by default on the global welfare of the animals. For that, there is no one solution, but there are many solutions, some are directly affecting us and our animals and the others are affecting indirectly. If the concept of balanced life is well understood, many of our problems will be solved.

The fasting carries deeper meanings than some people imagine; it passes the limits of self-control to control the spirit. In our case that will mean, one will be happy and satisfied if he consumes only his demand for food. One will do such practice in a relaxing form even with happiness. Perhaps this is possible for a human being who has a mind that thinks and chooses the correct choice. In wildlife animals, some species are able to limit their demands. But for domesticated animals that have found no one but a human as a teacher, they eat everything that is offered to them. So, there is no chance that they will reach the best choice, only human can do that for them. But did he do? Therefore, the chickens do not stop eating day or night as long as they see food in front of them. The last example indicates a deviation from the inevitability of life's balance (Amara 2013).

As for animal husbandry, in a quick overview, one can say that we have deviated from the inevitability of a balanced life. Perhaps because we are deeply immersed in many practices, we may need a simple person who does not know what we are doing to determine for us what is normal and acceptable and what is abnormal and reprehensible. For example, if we brought someone who is not aware of our modern practice and asked him to mix some animal products to put in the cows' food, he would probably say: "The cows will not eat this food." At the beginning of the appearance of crowded fast-farmed chickens, many people with their natural sense refused to eat them because they found that a 45-day-old chicken could not be with this large size and weight if it grew in natural conditions and that their bones were fragile and did not match those of similar sizes that grew naturally. In fact, the person who raised them and sold them after 45 days kept the amount allocated for his personal consumption for several additional weeks was to have the same sense, so he fed them natural food and put them with what is called local chickens (slow open traditional farming) where life is normal. After growing them for enough days, he is now satisfied with their quality. In fact, this practice was widely used in the first appearance of the poultry of the fast crowded farm (in Egypt and might be in other places) but over time it has disappeared. With other animals, many lay people spontaneously said that the cause of mad cow disease was due to feeding them animal components, and the research later proved that to be true and correct.

10.5 THE ANIMAL WELFARE

10.5.1 Interact with the Best Animals That Could Accompany Us!

Animal welfare is described historically by many who are related to different civilizations. Even animals are subjected to the mummification practices to preserve their corpses in many countries, including Egypt. No one could ignore that the Indians have a specific belief and philosophy about animals. Some believe that the souls of dead persons could come again in the form of animals. Correct or incorrect, but the fact is that the animals are creatures that have souls, have family, and can communicate. In clear words, animals are groups and nations like us. In some religions, they are more intelligent than us because they refuse to be minded and have the clear ability to select the correct way or the incorrect way. They select to be as they are and that only human accepts to have a mind, so God will reclaim him but not them and others (e.g., mountains, plants, etc., in the concept that all are creatures). For that, animals are respected in different places, and the best of them are selected to accompany humans. The belief that horses have honor leads humans to give them extra respect. The hidden test for the purity of the blood of the Arabic horse is that it mates only within houses, does not drag, and does not eat from the food you offer, if you hit it. Elephants, camels, lamas, and others have similar characteristics. Some belive that characters could be transferred from animals to humans by eating or accompanying! Camels, which are tough animals, will be considered by knowledgeable humans as a remark for some behaviors of persons who are on daily contact with them. Pig is not eaten by some ethnic groups, due to a belief that it will negatively affect the human characters and morals. Generally, and collectively, humans have succeeded over the ages to select the best animals that all the family members could interact with. In harsh conditions like in the open desert, the choices will be different due to other factors.

Most of those described animals' behaviors and their effect on humans are documented with oral or written stories.

10.5.2 The Animal Welfare in Its Old and Modern Versions

The interest in animals is clear in many old civilizations such as the cats in the Egyptian old civilization; the elephants, cows, etc., in Indian civilization; the Banda in Chinese civilization; and the Eagle in the Roman civilization. Some interest is extended from one civilization to another. The western civilization in a military point of view gains excessive success after the use of Arabic horses or breed its own horses with selected Arabic horses. Stories tell that those Arabic horses were crying when they were taken from their original places. The welfare of the Arabic horses is clear in their literature, stories, practices, etc. The horses have names, family names, family trees, and the like.

Meanwhile, the modern animal welfare as a discipline started with the publication of the Rogers-Brambell (a report) on the welfare of farm animals, issued by the British Government in 1965 (Rogers-Brambell 1965; Carenzi and Verga 2009). It is clear that there is an intelligent eye sense that the animal is suffering. The response comes to put some regulation on how animals are treated by humans and that it must be a correction. Early reports pointed out that factory farming is threatening the family farm (traditional farming style) with extinction (Fraser 2001; Dolan and Jr. 1986) (Figure 10.1). Other studies are concerned with the consumer's opinions and beliefs (Taylor and Signal 2009). Reports describe that consumers are willing to pay more for products that come from animals that are grown properly in proper welfare (Taylor and Signal 2009). Meanwhile, for many scientists, animal welfare is defined and measured on the basis of how well the animal is performing in a biological functioning perspective (Barnett and Hemsworth 2009). For others, the animal preferences are influenced either by the animal's emotions or by the animal's important biological requirements (Barnett and Hemsworth 2009). Animal welfare is proposed to be taught in schools (including wild animals) (Hawkins and Williams 2017).

One should note the huge number of differences between the family farm where animals are in daily observation and in direct contact with their owners and those in the bulk crowded production where many owners see them as money. For that, many try to protect their business by any means. Robbins (1987) highlighted that in the modern agribusiness, producers seek profit without reference to a documented ethical sensitivity to the animals (Fraser 2001; Robbins 1987). In fact, even most writers tend to use delicate words, but many scientists instead use sharp words when they are writing about animal welfare. That might be explained by their hope to support animals against the bad conditions by improving their welfare. And for that, they should not be reclaimed by any.

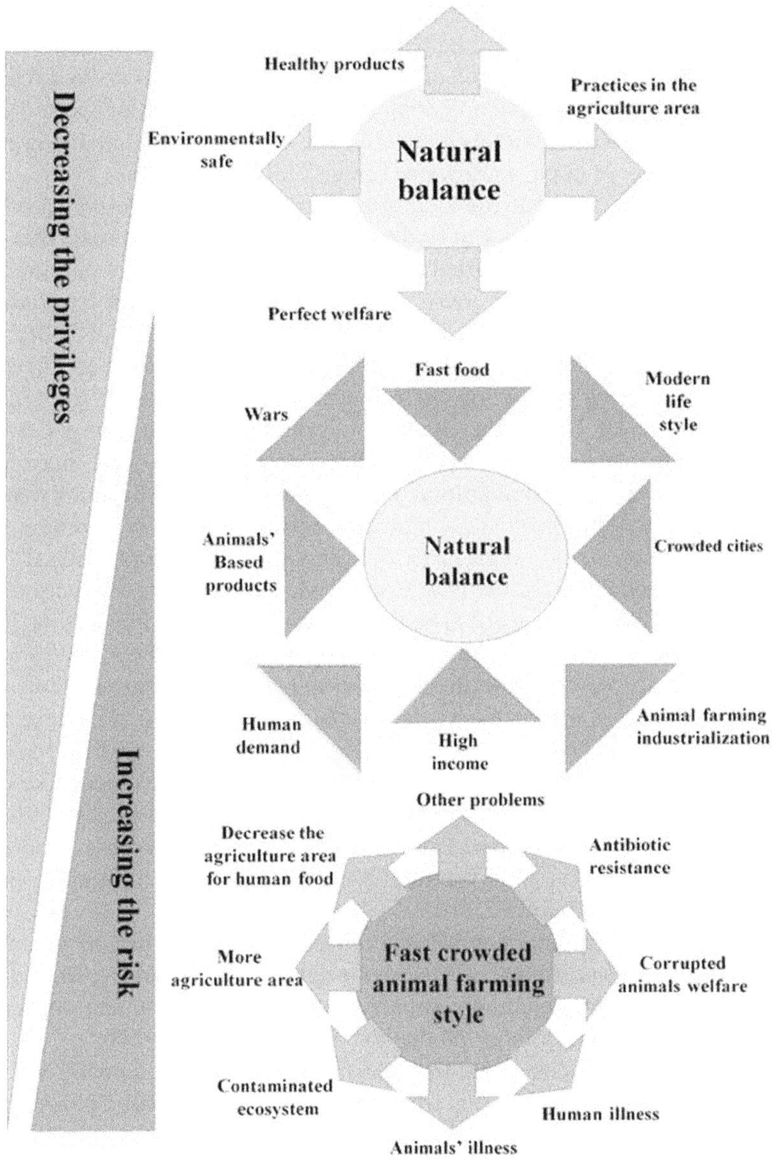

FIGURE 10.1 The figure represents the differences between the natural balance in the family farm, which was the main farming style that ran overall the world for hundreds of years, and the modern fast crowded animal farming style. The family farm is considered to be well balanced where animal welfare is well respected. The image shows the privileges of the family farm in the traditional style and the changes that affect it like wars, crowded cities, modern lifestyle, the increase in human demand, animal farming industrialization, etc. All such changes lead to an increase in the use of the fast crowded animal farming style with its all side effects such as the corruption of the animal welfare, animals' illness, human illness, antibiotic resistance, consumes the agriculture area for producing food to feed the animals in the crowded farming style farms and industries, etc. (generated in this study).

Another angle of the modern farming style is the side effect on the amount of cultivated area that is used to feed humans. Now, the farming business has in many places the priority over the allocated amount of area for human food. As a result, the number of hungry individuals is increasing every year. Coats (1989) reported that the vast tonnage of food fed to animals to produce the heavy meat-based diet is given at the expense of hungry people worldwide (Fraser 2001; Coats 1989).

Questions are asked about the side effects of the modern animal production style on the animal. Some lines can be allocated using some key points that can be considered as a title of other subtitled one, for example, the welfare of animals and animal producers, the health of consumers, the sustainability of food production, the global environment, and world hunger and food security (Fraser 2001). Fraser (2001) put the previous headlines as a point to help analyze the welfare of the animals in certain farming production environment, or in other words "from the farm to the consumers." He linked the change in agriculture to the changes that happened in the culture and *vice versa* (Fraser 2001). Other studies were more specific and focused only on one type of animal. For example, the dairy buffalos were studied to analyze specific indicators including excessive thinning or fattening; cleanliness; health status; social, aggressive, oral abnormal behaviors; animal–human relationship; and housing factors (De Rosa et al. 2009). There are many factors that could be addressed. Some could be generalized and others are specific like the odor emissions in broiler production (Sharma et al. 2017). Others like to analyze one factor and trace its effect on the animals. For example, Frazzoli et al. (2015) used trace elements as a monitoring parameter in animal feed (Frazzoli, Bocca, and Mantovani 2015). Trace elements are essential for normal healthy body functioning (Frazzoli, Bocca, and Mantovani 2015). Or a single element like copper metabolism (Leeson 2009), dietary glycine supplementation (Akinde 2014), and antioxidant defense system (Yigit, Panda, and Cherian 2014) are also considered. For more details about the antioxidant and the philosophy behind using them in balanced way, refer to Amara (2010).

Other studies use a comparison between different strategies for one treatment as a tool to understand the different effects on the animals. For example, Bloksma et al (2008) reported a study including comparisons between organic and conventional production processes (e.g., in milk production) (Bloksma et al. 2008).

Blokhuis et al. (2010) concluded that specialists in a certain domain will not be enough to cover all aspects of animal welfare (Blokhuis et al. 2010). The most appropriate specialist expertise in the multidisciplinary field of animal welfare should be brought to optimize the animals' welfare conditions (Blokhuis et al. 2010; Millman 2009). Perhaps one expert has been forgotten or neglected, he is the farmer who could be represented in the family farming or the traditional farmer who has a house, a farm, and few animals to satisfy his family's demand with a small business to be able to cover the cost of other life's demands.

Canali and Keeling (2009) made animal welfare simpler in a few words: good feeding, good housing, good health, and appropriate behavior (Canali and Keeling 2009). Animal welfare or "quality of life" represents protecting animals from thirst, hunger and malnutrition, discomfort, pain, injury, fear and distress, and disease, as well as expressing normal behavior.

Green and Mellor (2011) described animal welfare as "quality of life" that is concerned with nutritional, environmental, health, behavioral, and mental factors and that animals must be protected from thirst, hunger and malnutrition, discomfort, pain, injury, and disease distress.

10.6 REMARKABLE ACTIVITIES CHANGE NATURAL BALANCE AND PROPOSED SOLUTIONS

To summarize some remarkable activities that prove our deviation from the natural practices, some selected points could be spotted. They represent examples concerning animal feeding, growing, treatment, welfare, etc.

10.6.1 The Slaughtering of Newborn Buffalo

The milk of newly born animals is taken from them and they are given synthetic milk. In best practices, large quantities are taken and the animals are supplied with the rest. Unfortunately, in particular, the young buffalo is slaughtered after few months (based on its size) to take all its mother milk. In fact, some farmers do that with self-pain because they are in need of money to satisfy their family members' demands. On the other hand, the meat of the young buffalo is considered to be digestible for old people and persons with certain illnesses. The government must support the farmers to raise their newly born buffalo and could in the future get a percentage of its sealing money. Regulations must be applied to guarantee their rights. Slaughtering newly born or young buffalo must be stopped.

10.6.2 Bovine Spongiform Encephalopathy (BSE) or Mad Cow Disease

The animals are given foods that might contain animal and fish products, and in some cases hormones, to increase the amount of milk and meat produced. Farm animals are herbivorous and animal and fish products should be avoided and hormones should be used correctly or avoided completely. BSE, or mad cow disease, is an incurable and invariably fatal neurodegenerative disease of cattle. Symptoms include abnormal behavior, trouble walking, and weight loss (Casalone and Hope 2018). Later, the cow loses its ability to function normally (Casalone and Hope 2018). The time between infection and onset of symptoms is supposed to be approximately 4–5 years (Nathanson, Wilesmith, and Griot 1997). Time from onset of symptoms to death is generally weeks to months (Nathanson, Wilesmith, and Griot 1997). BSE is thought to be due to an infection by a misfolded protein, known as a prion (Nathanson, Wilesmith, and Griot 1997), which comes through the feeding with a meat-and-bone meal (Na 2001). The outbreak increased throughout the UK due to the practice of feeding meat-and-bone meal to young calves of dairy cows (Nathanson, Wilesmith, and Griot 1997). Cases are suspected based on symptoms and confirmed by examination of the brain (Casalone and Hope 2018). Cases are classified as classic or atypical, with the latter divided into H- and L types (Casalone and Hope 2018).

It is a type of transmissible spongiform encephalopathy. The solution will be simple: letting young calves feed on their mothers' (cows) milk.

10.6.3 The Farming Spaces

The free space given to the farm animals either mammalian, poultry, or fishes should be considered as a major part in the success of any animal farming process. In many cases, the farming places are not suitable for correct and healthy growing conditions. The crowded farming style has many negative factors described in different parts of this chapter. Animals, birds, and fish should be given a respective space that can guarantee better farming, growing conditions, and better products. Antibiotics and other antimicrobial compounds are given to cover the crowded negative condition as well as any weakness or immunocompromised conditions due to exhausting the animals to produce an extra amount of milk, or giving extra-a number of foods that exhaust their organs. Animals in many cases are not subjected to correct day and night. Animals should live normal day/night life, eat enough but not extra food, avoid condition could not be covered but only if antibiotics or antimicrobials are used. The animal crowding farming style is dominant for growing different types of animals including animals, fish, poultry, etc. It can be considered as a common wrong practice in modern animal farming. Animals should be given correct spaces that are healthy and suitable for animal welfare. The farming places are usually wet and contaminated with animal wastes. Besides the correct space, the space condition should be respected and be clean and healthy. In case of poultry growing conditions such as the day/night, cages, antibiotics, light, hormones, feed style and contents, growing conditions and the like affect the animal welfare and the quality of the final products. For more details about the poultry farming, refer to Amara et al (2024).

10.6.4 Crowded Fish Farming

One example could be given about the deviation from the natural balance is the crowded fish production worldwide. In Chile (1973 and 1976), salmonids farming is one of the nation incomes (Lozano et al. 2018). Four species of salmonids were cultivated: Chinook (*Oncorhynchus tshawytscha*), rainbow trout (*Oncorhynchus mykiss*), Atlantic salmon (*Salmo salar*), and coho (*Oncorhynchus kisutch*). Between 1987 and 2010, diseases appeared in the salmon farming because of the ectoparasite Caligus and bacterial kidney disease in 1987 (Cabello 2004; Ibieta et al. 2011); the presence of *Piscirickettsia salmonis* in 1989 (Lozano et al. 2018); and the presence of high mortality in 1997. The outbreaks happened due to the infectious salmon anemia disease in the years 2002, 2007, and 2008 (Cabello 2004). Some authors claimed that this has happened due to the use of antibiotics in 1989 (Cabello 2004; Ibieta et al. 2011). Generally, in fish farming, hormones, ammonia level in water, microbes, and the crowded cultivation conditions affect the animal welfare. The uneaten food will contaminate the water and the eaten food will also come back to the water as waste. Antibiotics will accumulate in the aquaculture water system systematically. Some have reported similar conclusions early. For that, Meek et al. (2015)

suggested that about 70%–80% of antibiotics come from fish excreta (Meek, Vyas, and Piddock 2015; Serrano 2005; Burridge et al. 2010).

10.6.5 Control the Use of Antibiotics and Other Medications

Secondary problems are elevated from time to time as a result of the bad practices including antibiotic resistance (Wang et al. 2024; Davies and Davies 2010; Miller 2016; Moudgil et al. 2017), antimicrobial resistance (Chantziaras et al. 2013; Grace 2015; Sharma et al. 2018; Xiong, Sun, and Zeng 2018), epidemic infections due to the growing conditions, new diseases like mad cow disease, new viruses, and the like. Other technical problems might not be related to the growing conditions but to other factors that might directly or indirectly affect the animals like the production costs, utility, marketing, the economy of the country, the labor cost, and the man/hour. Other problems are related to storage, home storage, cooking, additives to food and the like. For more details about the antibiotic use, the antibiotic alternative, the antibiotic resistance, etc., refer to Amara et al. (2024).

10.6.6 The End-Product Processing

Proper farming conditions include animals feed, live, sleep, move, drink, etc. But if animals are subjected to fear that will lead to increase adrenaline hormone secretion in their bodies. The slaughtering is a particular event where an animal feels a fear. Avoiding such consequences and the other ones that might affect negatively on the animal end-product will be helpful. This thing will let us protecting the animal from many of the problems they might face like the microbial load, the drug restudies, the skin quality (linked to clean and dry environment), the amount of the hormones, and the like. That's mostly because the end product will be consumed by us. After the slaughtering process, the story will not end yet. The steps included in preparing the end product to reach the market might interfere with the product quality. The contamination with microbes; chemical residues such as the preservative, contaminated salt with heavy elements, and other chemicals; and the like; all collectively must be controlled and avoided. It is the role of the manufacturer's quality control to guarantee the quality of the animals and products. One might sense that this part is concerned with human health but not with animal welfare. In reality, it is also concerned with animal welfare because these practices will protect the laborers who are working in the animal farming industrial sector either those who are working in the final or the end-product processing and could be become a built-in source of contamination, or those labors who are working in the regular works and in contact with the animals. Or their relative, friends, neighboring, etc., who should be in contact with them at any time. The image is more collective than being personalized. The protection strategies must protect all the components that in fact will include every unit in the earth's ecosystem. For that, pesticides, synthetic fertilizers, drug residues, hormones, heavy metals, microbes, toxins, and the like are invited candidates.

10.7 THE ACCUMULATION OF DIFFERENT MISTAKES AND THEIR EFFECT ON THE ANIMAL WELFARE AND THE HUMAN HEALTH

All the above-described accumulated mistakes will finally effect on human health either directly or indirectly when humans consume animal products. The most important accumulated forms include different microbial species, different resistance microbes harboring antibiotic resistance genes, antibiotic residues, drug residues, and different hormones, including adrenaline during the unwise killing process, extra fat content, and the like.

It is important to highlight that some factors are linked to each other and lead to the accumulation of certain problems. Cutting the first factor will cut all subsequent expected problems. For example, the crowded animals in the crowded farming style led to speed the transfer of the pathogens, use more antibiotics, lead to the elevation of the antibiotic resistance microbes, increase contact with the animal wastes will increase the spread of the infections, incorrect animal movement will cause many physical and physiological diseases, etc. If animals are farming in a respective condition and correct farming space, aerated, dry, clean, etc., that will decrease the pathogen transmission, reduce illness, reduce the use of antibiotics, reduce the contact with the animal wastes, and give it a chance to dry faster, keep the animal moving freely and in relaxed mode. Correct one factor will correct all its related consequences.

10.7.1 The Demand of the Animals, the Producer and the Consumers: All Are Linked Positively or Negatively

To apply the concept of the inevitability of balanced life, one should consider the demand of the animal itself and not consider the demand of the producer or the consumers. The animal mother's milk is basically produced to feed the newborn. And it is usually exceeding the newborn demand and enables humans to get the benefit from the rest. Normal farmers give the newborn chance to get enough milk from its mother, an amount which is enough to support a correct healthy (for the mother and for the newborn) and to express normal behavior. Being lactated from his mother, in direct contact with it and under its kindness and recognition, etc., the newborn will grow normally, happy, healthy, etc. being able to eat and search for food, its mother will keep giving milk for extra-long time nearly with the same amount till getting pregnant again.

Another angle for applying the concept of the inevitability of balanced life is simply applying the normal day and night. For example, in case of poultry production, using extended day or 24h lighting conditions will exhaust the animals. It is a fraud way to keep it eats for a long time which will give additional weight but will give a weak unhealthy and abnormal animal, unable to move correctly, etc.

10.7.2 The Antibiotic Resistance as a Specific Example about the Misuse of the Nature Gifts

The use of antimicrobial agents including antibiotics is an interesting issue for both humans and animals. It is selective for the prokaryotic microbes, and in correct dosage/correct time has minor side effects on the body. But it could not select between beneficial and harmful microbes. In addition, it could not kill microbes that are protected in places where the proper concentration of antibiotics does not reach the microbes or microbes that form biofilm. One type of microbes under the used concentration could survive, they are those who have antibiotic resistance genes or antibiotic resistance mechanisms, or could develop a resistance mechanism or acquire a resistance gene in the correct time. Moreover, they could transfer their mobile elements to other microbes to turn them from sensitive to the antibiotic(s) to be resistant. Finally, sensitive microbes will be dead, resistance will survive, and then, resistance one will dominate the treated environment and a new antibiotic should be used. The antibiotic misuse did not harm only the human and animals but also the environment. The different antibiotics have done nearly the same described mechanism (as above) and finally resistance microbes become dominant. They transferred then widely in new victims and new habitats due to the natural movement of water, air, soil, etc. Or transfer by force through the traveling of humans and animals from the infected area to other new and clean one. Of course, and due to the immunity of the new community, the newly \arrived microbes will be more virulent in their new environment. For that, one could find suddenly a disease that appears in certain places but not in all places. Human has been spotted as the main responsible one for the animals' growth. He must be aware with all those issues. Those who are working in animal farming, particularly in poultry production, must have some training and some knowledge of basic microbiology. It is not enough to disinfect the hands and the clothes of the worker and cover their heads to be safe from the pathogen transfer. Because it is a fact that those practices are not enough, and after many trails to make them enough, finally, the antibiotics come again as a solver for this problem, where animals are given antibiotics in their fight to survive during their pouring 45 eating and growing days. Did the story end at this point. Nothing is ended, the antibiotics come out as a contaminant in the animal excreta and discharge, and contaminate the surrounding environment, the water, the landscape, the cultivating area, the lakes, the rivers, etc. Yes, the condition of the hospitals guarantees a relaxed place for the pathogenic antibiotic-resistant microbes, but by any means, the area of the hospitals could not be compared to the total area of places that practice animal farming and use antibiotics as a basic component to allow animals to survive. Moreover, the antibiotics are given only to the patients who have stopped using them after recovery, but the antibiotics given to the farming animals reach nearly every house around the world, except those who are not linked to the modern system for some purposes.

How did humans interact with this clear problem, simply if resistance to an antibiotic is elevated? He searches to isolate/design/produce new antibiotics! For more details about this issue, kindly refer to Amara et al. (2024). The newest one becomes useless after some time due to the main problem: the antibiotic resistance. This problem extends from one generation to another. It needs deeper analysis for the basis of the problem rather than fast solutions. The antibiotics' misuse led to antibiotic resistance that reached 100% resistance to all existing antibiotics in the market. One example is the MRSA. There are huge numbers of antimicrobial agents but few are using them as an alternative. To analyze this phenomenon, one should go back before the antibiotic was purified and how humans have used other forms of antimicrobial agents (El-Baky and Al Fattah Amara 2023) or other practices (traditional slow animal farming style) that did not need the use of antimicrobial agents, including antibiotics. For that, it must be clear that the problem is not only in the antibiotics and the pathogen interaction, but in the running scenarios for farming animals or even for treating animals and humans. Other issues are also addressed, such as the role of the immune system in reducing the dependence on the antimicrobial agents. Antibiotics fight microbes but keep the immune system on vacation. The infection with a virulence pathogen in an unexperienced immune system might be fatal. Because antibiotics are the easy way to fight microbes, their use becomes common, particularly in the medicinal and veterinary systems at first. The use of antibiotics in agriculture, aquaculture, food industry, etc., led to the omission of the other forms of antimicrobial agents from the competition. Antibiotics are becoming widespread, and microbial biodiversity is changing from the human and animal microflora (Afrc 1989), which contains probiotics, to the open natural ecosystem. The load of the antibiotics and their residues – that contaminate different ecosystems; that encourage microbes to transfer genes either through the common gene transformation roles (e.g., conjugation and transduction) or by newly discovered ways such as the ability of dead microbes to transform their resistance genes to viable ones (Amara et al. 2024). In general, simply the nature balance was changed. Different factors are included. Even the type of food that we eat could affect the type of our microflora and there exist genes.

10.8 WHY NATURE, BEFORE THE ERA OF ANTIBIOTICS, WAS BALANCED AND HEALTHY?

The answer to this question will let us able to survive. It is by reducing the amount of the antibiotic residues that contaminate nature, controlling the use of the antibiotics, and practicing effective roles that help us to get the maximum benefit from the antibiotic without or with minimum problems. Omitting the condition that forces us to use the antibiotics is a critical point. For example, in crowded animal farming, there is no choice but one must use antibiotics. But historically, our family farming style succeeded for thousands of years in growing healthy animals without the use of antibiotics or even the use of any form of antimicrobial agent. In other words, the presence of the antibiotics is a part of the problem and not only a solution that has been misused (Figure 10.1). It is time to correct this problem forever. But this needs a long-term plan to build a successful farming process that does not depend only on the antibiotics. Some idea could be addressed such as (1) using all the available safe alternative to the antibiotics; (2) stopping any bulk use for the antibiotic in any sector and restrict them to be used only in disease treatment, or for protecting human with specific illness (e.g., immunocompromised patients); (3) using the traditional slow farming style (TSFS) model or some name it family farming side by side with improving the crowded fast farming style (CFFS) production model in the direction of using fewer antibiotics; (4) reducing the use of farming conditions that lead to the use of antibiotics like cages, crowded animals, long lighting systems and the like; treating the agriculture wastes and stop the antibiotic residues from reaching the nature; (5) giving open nature areas that are contaminated by antibiotics enough time to get rid of the antibiotic residues such as antibiotic residues that are degraded by hydrocarbon degrading microbes, and then enough time to cure the undesirable genes such as the R-Factor (plasmid curing); (6) keeping a suitable number of antibiotics out of the market to be used only in emergence cases; (7) installing perfect surveillance system to monitor the antibiotic residue, AR genes, and expected sources for the antibiotics' residues (hospitals, industries, etc.); (8) giving natural immunity to human and animal a space and reduce the improper use of antibiotics; (9) correcting the public and the farmer understand about the safety of using antibiotics in food products; and (10) educating the students a correct image about how antibiotic resistance microbes could be happened to avoid its happening.

10.9 THE SIDE EFFECT OF ANTIBIOTIC RESIDUES ON THE MICROBIAL BIODIVERSITY

The effect of antibiotics on microbial biodiversity is not separated from the points discussed above, but it has been given more space due to its importance (Hussein and Amara 2006). It has been written specifically to omit any incorrect idea about the microbes, which in fact are one of the wonderful and helpful organisms to humans if one did not consider the few countable pathogenic microbes. Generally, and collectively, microbes are not all evil. And pathogenic microbes are minute in their number and existence. But perhaps because they are virulent and have many virulent factors, they could, if succeeded, bypass the immune system of a healthy person and they could harm a lot. Meanwhile due to their few numbers in a balanced nature ecosystem, personal hygiene might be enough to protect us against them or at least most of them and to keep them away. Such hygiene can be achieved only by cleaning bodies or other tissues with water and keeping them dry. For that, hand (and bodies) washing is an essential part of surviving against microbes. Apparently, microbes are mainly created to clean nature from different wastes, including antibiotics. They do this activity in the desert, cities, farms, forests, marine ecosystem, and nearly every place. Wastes are food for microbes. A simple observation shows that using disinfectant to clean walls will keep them clean for some time, but after that the places that were cleaned with disinfectants will show fungus growth. But walls that were cleaned with water (only) and subjected to enough drying did not show any fungus growth. Even in our bodies, the parts of the skin that are subjected to moisture longer show microbial infections. Even the human body is a food for microbes but when this food is protected by dry skin, active immune systems, etc., microbial trials will be blocked. Nonetheless, when the human immune system is compromised, attacks start to appear, which is an early signal about the weakness of the immune system because of internal or external factors. Did we respond to these signals correctly? Pathogenic microbes are living in nature with many other microbes, those which fix nitrogen, degrade cellulose, degrade hydrocarbons, degrade protein, degrade the corpses of dead animals and human, and clean the nature of many contaminants. They all have the power to survive in the balanced life under normal conditions. But not all of them have antibiotic-resistant genes (ARGs) (Amara and Shibl 2015). Exposing the nature to antibiotic residue will kill them if they do not acquire resistance genes or build a successful resistance mechanism. In conclusion, we enrich the nature by factors that shift its balance toward favoring the condition of the ARMs.

10.10 WE SHOULD START TO TAKE ACTION

After the above-discussed points, examples, etc., one could summarize why we should rethink our way of growing farming animals. Those points could be summarized as follows:

1. To stop the critical need for using antibiotics, or in other words, we should use antibiotics only for treatment and not for protection.
2. In the absence of the use of antibiotics as a protector (prophylaxis), we will force ourselves to find other alternative and other solutions that do not elevate antibiotics or drug-resistant microbes.
3. To save ourselves, the other, the animals, the nature, the microbial biodiversity, and every component that could be affected directly or indirectly by the negative points addressed due to the bulk CFFS.
4. Using the concept of the inevitability of balanced life will let us navigate through the bad and the correct options, understand the nature balance, maintain the animals' welfare, protect ourselves, protect the animals, protect the nature, and the like.
5. The money we gain from the modern farming style will be paid in treating illness, infections with newly born resistant microbes, changes in the nature biodiversity, and the like.
6. The family farming style should be preserved, encouraged, financially supported, and documented. Their product must be sealed with higher price to encourage the family farmers to survive and to stop the distraction happened to them due to the bulk crowded production (Taylor and Signal 2009).
7. The family farming style could be partially enlarged to be more profitable without changing its main beneficial components.
8. Some serious indirect consequences can occur due to the deviation from the inevitability of balanced life like killing important hydrocarbon-degradable microbes that eliminate hydrocarbon wastes from the nature including antibiotics.
9. Putting a complex process like the CFFS in the hands of uneducated persons might be catastrophic. The education system must correctly share the knowledge concerning animal farming, their welfare, safety, biodiversity, antibiotic use, etc., with everyone who is going to grow farm animals either using the family farming style or any of the modern ones.
10. Collaborative scientists, experts (Barnett and Hemsworth 2009), workers, managers, businessmen (Lange et al. 2020; Truong et al. 2021), etc., must be gathered to put a clear plan for the best conditions for animal farming.
11. The family farming style must get more financial support, governmental support, etc. And the area where it is located must not be distracted or highly automated. That is because (as described above), if the animals become useless, there will be no selection for the best animals that could support the farmer. The farmers can then be rewarded by increasing the price of their products.

12. Everyone must respect the animals even before they are born, during their birth, during their lactation period, maturation, etc., and even during their slaughtering.
13. Thinkers, politicians, teachers, film producers, etc., should think about how to integrate the animal welfare issues in the minds of the public and do a simple way to introduce the concept of the balanced life to the public. The concept of that, 'if I am safe to that enough', must be omitted from our societies because any corruption could harm us in the end.
14. All the components of the earth should be handled wisely and respectfully to protect the natural balance, the resources, the running biological cycles, and the like.

This chapter could not address all the ideas about how we can collectively maintain, build, or rebuild a balanced life, but it only gives a hint about where we can find our correct way and how we could start. In fact, the issue is so great to be summarized in one chapter or even in one book.

10.11 CONCLUSION

The inevitability of balanced life is a concept that interested us to react correctly with nature components. The chapter addresses examples of how changes in nature negatively affect such a balance. One could not say that it could affect positively because mostly the nature was already in balance. So, changes will cause disturbance in such a balance. The nature could tolerate some changes, but within certain limits and for certain times. One minor effect could have a large consequence. By understanding the major points that govern the natural balance, one could interact correctly and build a correct model for nearly everything, including animal farming, animal welfare, correct protection, correct treatment against pathogenic microbes, etc. The inevitability of balanced life did not interfere with the economy of the animal farming sector, but it corrects some critical deviations that harm animals and, by default, harm the human, the nature, the environment, etc. Our bodies, the animals' bodies, the nature and nearly everything, each has enough power to survive and to leave correctly side by side with the other components if they react spontaneously and take their need only from the other and from the nature. The traditional ancient farming style or what is called the family farming has succeeded over the ages and should be kept to survive for the best of all of us and the coming generations. It must be a concept to avoid eradicating a system proven success because we believe that their new one or more than one is profitable. If our mistakes are not corrected, the nature will correct them. Running the TSFS, using alternatives to the antibiotics, respecting the animal welfare, respecting the nature, understanding the inevitability of balanced life, etc., all collectively will guarantee healthy animals, proper welfare, safe animals' products that do not harm any of the nature components either in the long or the short run, high end-product quality, etc. The concept of the animal welfare, the elevation of the antibiotic resistance, the importance of the family farming style, and the other idea, models, explanations, phenomenon that are described in this chapter should be clear and available for everyone in concern with the addressed issues, particularly the animal farming and the animal welfare, and can be educated to increase the global awareness of the issues addressed in this chapter (Figure 10.1). The subject of the balanced life, the inevitability of balanced life, the animal welfare, etc., could not be discussed collectively in one chapter. However, some important key issues are highlighted and addressed.

REFERENCES

Afrc, R. Fuller. 1989. "Probiotics in man and animals." *Journal of Applied Bacteriology* 66 (5):365–378. doi:10.1111/j.1365-2672.1989.tb05105.x.

Akinde, D. O. 2014. "Amino acid efficiency with dietary glycine supplementation: Part 1." *World's Poultry Science Journal* 70 (3):461–474. doi:10.1017/S004393391400052X.

Amara, A. A., N. Abd El-Baky, M. M. Salem-Bekhit, K. M. A. Zoheir, and M. M. M. Ahmed. 2024. Antibiotic resistance from theory to facts. In: A. Gupta and R. Prasad (Eds.), *Antimicrobial Resistance in Agriculture and its* Consequences (pp. 1–26). CRC Press.

Amara, A. A. A. F. 2010. "The philosophy behind exo/endo/existing antioxidants and our built-in oxidant and antioxidant system." *Pharmazie* 65 (10):711–719. doi:10.1691/ph.2010.0557.

Amara, A. A. A. F. 2013. "The inevitability of balanced lives: Genes and, foods in action and, interactions." *IIOAB Journal* 4 (2):1–27.

Amara, A. A. A. F., and A. M. Shibl. 2015. "A hypothesis about the variation within B-lactamases: An epigenetic-like mechanism." *IIOAB Journal* 6 (1):1–11.

Amara, A. A., A. J. Neama, A. Hussein, E. A. Hashish, and S. A. Sheweita. 2014. "Evaluation the surface antigen of the *Salmonella typhimurium* ATCC 14028 ghosts prepared by "sLRP". *The Scientific World Journal*. doi:10.1155/2014/840863.

Amara, A. A., M. M. Salem-Bekhit, and F. K. Alanazi. 2013. "Sponge-like: A new protocol for preparing bacterial ghosts." *The Scientific World Journal*. doi:10.1155/2013/545741.

Amara, A. A., A. M. Shibl, and A-K. F. Tawfik. 2012. "Deeper in the antibiotics paradigm: Microbes react individually and collectively for spreading the resistance." *African Journal of Microbiology Research* 6 (9):2010–2019.

Arieli, A., J. E. Vallimont, Y. Aharoni, and G. A. Varga. 2001. "Monensin and growth hormone effects on glucose metabolism in the prepartum cow." *Journal of Dairy Science* 84 (12):2770–2776. doi:10.3168/jds.S0022-0302(01)74731-5.

Barnett, J. L., and P. H. Hemsworth. 2009. "Welfare monitoring schemes: Using research to safeguard welfare of animals on the farm." *Journal of Applied Animal Welfare Science* 12 (2):114–131. doi:10.1080/10888700902719856.

Blokhuis, H. J., I. Veissier, M. Miele, and B. Jones. 2010. "The Welfare Quality® project and beyond: Safeguarding farm animal well-being." *Acta Agriculturae Scandinavica, Section A — Animal Science* 60 (3):129–140. doi:10.1080/09064702.2010.523480.

Bloksma, J., R. Adriaansen-Tennekes, M. Huber, L. P. L. van de Vijver, T. Baars, and J. de Wit. 2008. "Comparison of organic and conventional raw milk quality in the Netherlands." *Biological Agriculture & Horticulture* 26 (1):69–83. doi:10.1080/01448765.2008.9755070.

Bourdichon, F., S. Casaregola, C. Farrokh, C. J. Frisvad, and L. M. Gerds. 2012. "Food fermentations: Microorganisms with technological beneficial use." *International Journal of Food Microbiology* 154:87–97.

Burridge, L., J. S. Weis, F. Cabello, J. Pizarro, and K. Bostick. 2010. "Chemical use in salmon aquaculture: A review of current practices and possible environmental effects." *Aquaculture* 306 (1–4):7–23. doi:10.1016/j.aquaculture.2010.05.020.

Cabello, F. C. 2004. "Antibióticos y acuicultura en Chile: Consecuencias para la salud humana y animal." *Revista médica de Chile* 132 (8). doi:10.4067/s0034-98872004000800014.

Canali, E., and L. Keeling. 2009. "Welfare Quality® project: From scientific research to on farm assessment of animal welfare." *Italian Journal of Animal Science* 8 (sup2):900–903. doi:10.4081/ijas.2009.s2.900.

Carenzi, C., and M. Verga. 2009. "Animal welfare: Review of the scientific concept and definition." *Italian Journal of Animal Science* 8 (sup1):21–30. doi:10.4081/ijas.2009.s1.21.

Casalone, C., and J. Hope. 2018. "Atypical and classic bovine spongiform encephalopathy." In: M. J. Aminoff, F. Boller, and D. F. Swaab (Eds.), *Handbook of Clinical* Neurology (pp. 187–199). Elsevier.

Chantziaras, I., F. Boyen, B. Callens, and J. Dewulf. 2013. "Correlation between veterinary antimicrobial use and antimicrobial resistance in food-producing animals: A report on seven countries." *Journal of Antimicrobial Chemotherapy* 69 (3):827–834. doi:10.1093/jac/dkt443.

Coats, C. D. 1989. *Old MacDonald's Factory Farm: The Myth of the Traditional Farm and the Shocking Truth about Animal Suffering in Today's Agribusiness.* Continuum.

Davies, J., and D. Davies. 2010. "Origins and evolution of antibiotic resistance." *Microbiology and Molecular Biology Reviews* 74 (3):417–433. doi:10.1128/MMBR.00016-10.

De Rosa, G., F. Grasso, C. Pacelli, F. Napolitano, and C. Winckler. 2009. "The welfare of dairy buffalo." *Italian Journal of Animal Science* 8 (sup1):103–116. doi:10.4081/ijas.2009.s1.103.

Dias, I., M. Laranjo, M. E. Potes, A. C. Agulheiro-Santos, S. Ricardo-Rodrigues, A. R. Fialho, J. Véstia, M. J. Fraqueza, M. Oliveira, and M. Elias. 2020. "Autochthonous starter cultures are able to reduce biogenic amines in a traditional Portuguese smoked fermented sausage." *Microorganisms* 8 (5). doi:10.3390/microorganisms8050686.

Dolan, E. F. Jr. 1986. *Animal Rights.* New York: Franklin Watts.

El-Baky, N. A., and A. A. Al Fattah Amara. 2023. "Natural antimicrobial therapeutic peptides: Milk lactoferricin and spirulina platensis peptides." In: D. C. Petrescu and M. Şandru (Eds.), *Antimicrobials in Pharmaceutical and Medicinal Research* (pp. 19–48). CRC Press.

Fraser, D. 2001. "Farm animal production: Changing agriculture in a changing culture." *Journal of Applied Animal Welfare Science* 4 (3):175–190. doi:10.1207/S15327604JAWS0403_02.

Frazzoli, C., B. Bocca, and A. Mantovani. 2015. "The one health perspective in trace elements biomonitoring." *Journal of Toxicology and Environmental Health, Part B* 18 (7–8):344–370. doi:10.1080/10937404.2015.1085473.

Grace, D. 2015. Review of evidence on antimicrobial resistance and animal agriculture in developing countries. *Evidence on Demand*, 1–67 (2015). doi:10.12774/eod_cr.june2015.graced.

Green, T. C., and D. J. Mellor. 2011. "Extending ideas about animal welfare assessment to include 'quality of life' and related concepts." *New Zealand Veterinary Journal* 59 (6):263–271. doi:10.1080/00480169.2011.610283.

Hawkins, R. D., and J. M. Williams. 2017. "Assessing effectiveness of a nonhuman animal welfare education program for primary school children." *Journal of Applied Animal Welfare Science* 20 (3):240–256. doi:10.1080/10888705.2017.1305272.

Hussein, M. Z., and A. A. Amara. 2006. "Case-by-case study using antibiotic-EDTA combination to control pseudomonas aeruginosa." *Pakistan Journal of Pharmaceutical Sciences* 19 (3):236–243.

Ibieta, P., V. Tapia, C. Venegas, M. Hausdorf, and H. Takle. 2011. "Chilean salmon farming on the horizon of sustainability: Review of the development of a highly intensive production, the ISA crisis and implemented actions to reconstruct a more sustainable aquaculture industry." In: L. E. Cruz-Suárez (Ed.), *Aquaculture and the Environment—A Shared Destiny* (pp. 207–246). InTech.

Kumar, N., V. Kumar, S. M. Waheed, and D. Pradhan. 2020. "Efficacy of reuterin and bacteriocins nisin and pediocin in the preservation of raw milk from dairy farms." *Food Technology and Biotechnology* 58 (4):359–369. doi:10.17113/ftb.58.04.20.6728.

Lange, L., K. O. Connor, S. Arason, U. Bundgård-Jørgensen, A. Canalis, D. Carrez, J. Gallagher, N. Gøtke, C. Huyghe, B. Jarry, P. Llorente, M. Marinova, L. O. Martins, P. Mengal, P. Paiano, C. Panoutsou, L. Rodrigues, D. B. Stengel, Y. van der Meer, and H. Vieira. 2020. "Developing a sustainable and circular bio-based economy in EU: By partnering across sectors, upscaling and using new knowledge faster, and for the benefit of climate, environment & biodiversity, and people & business." *Frontiers in Bioengineering and Biotechnology* 8:619066. doi:10.3389/fbioe.2020.619066.

Leeson, S. 2009. "Copper metabolism and dietary needs." *World's Poultry Science Journal* 65 (3):353–366. doi:10.1017/S0043933909000269.

Lingbeck, J. M., P. Cordero, C. A. O'Bryan, M. G. Johnson, S. C. Ricke, and P. G. Crandall. 2014. "Functionality of liquid smoke as an all-natural antimicrobial in food preservation." *Meat Science* 97 (2):197–206. doi:10.1016/j.meatsci.2014.02.003.

Lipan, L., B. Rusu, E. Sendra, F. Hernandez, L. Vazquez-Araujo, D. C. Vodnar, and A. A. Carbonell-Barrachina. 2020. "Spray drying and storage of probiotic-enriched almond milk: Probiotic survival and physicochemical properties." *Journal of the Science of Food and Agriculture* 100 (9):3697–3708. doi:10.1002/jsfa.10409.

Liu, N., Q. Zhu, X. Zeng, B. Yang, M. Liang, L. Deng, L. He, C. Liang, R. Zhang, and J. Zhou. 2018. "Optimization of processing conditions of Chinese smoke-cured bacon (Larou) with a new natural coating solution during storage period." *Korean Journal for Food Science of Animal Resources* 38 (3):636–652. doi:10.5851/kosfa.2018.38.3.636.

Lozano, I., N. F. Díaz, S. Muñoz, and C. Riquelme. 2018. "Antibiotics in Chilean aquaculture: A review." In: R. S. Singh, M. Kumar, and J. Singh (Eds.), *Antibiotic Use in Animals* (pp. 57–80). InTech.

Lück, E. 1985. Chemical preservation of food." *Zentralblatt für Bakteriologie, Mikrobiologie und Hygiene. Series B: Environmental Microbiology* 180 (2-3):311–318.

Margalho, L. P., G. P. Jorge, D. A. P. Noleto, C. E. Silva, J. S. Abreu, M. V. F. Piran, M. Brocchi, and A. S. Sant'Ana. 2021. "Biopreservation and probiotic potential of a large set of lactic acid bacteria isolated from Brazilian artisanal cheeses: From screening to in product approach." *Microbiological Research* 242:126622.

Meek, R. W., H. Vyas, and L. J. V. Piddock. 2015. "Nonmedical uses of antibiotics: Time to restrict their use?" *PLoS Biology* 13 (10):e1002266. doi:10.1371/journal.pbio.1002266.

Miller, S. I. 2016. "Antibiotic resistance and regulation of the gram-negative bacterial outer membrane barrier by host innate immune molecules." *mBio* 7 (5):e01541–e01516. doi:10.1128/mBio.01541-16.

Millman, S. T. 2009. "Animal welfare—scientific approaches to the issues." *Journal of Applied Animal Welfare Science* 12 (2):88–96. doi:10.1080/10888700902719591.

Moudgil, P., J. S. Bedi, A. D. Moudgil, J. P. S. Gill, and R. S. Aulakh. 2017. "Emerging issue of antibiotic resistance from food producing animals in India: Perspective and legal framework." *Food Reviews International* 34 (5):447–462. doi:10.1080/87559129.2017.1326934.

Murphy, D., A. Ricci, Z. Auce, J. G. Beechinor, H. Bergendahl, R. Breathnach, J. Bureš, J. P. Duarte Da Silva, J. Hederová, P. Hekman, C. Ibrahim, E. Kozhuharov, G. Kulcsár, E. Lander Persson, J. M. Lenhardsson, P. Mačiulskis, I. Malemis, L. Markus-Cizelj, A. Michaelidou-Patsia, M. Nevalainen, P. Pasquali, J. Rouby, J. Schefferlie, W. Schlumbohm, M. Schmit, S. Spiteri, S. Srčič, L. Taban, T. Tiirats, B. Urbain, E. Vestergaard, A. Wachnik-Święcicka, J. Weeks, B. Zemann, A. Allende, D. Bolton, M. Chemaly, P. S. Fernandez Escamez, R. Girones, L. Herman, K. Koutsoumanis, R. Lindqvist, B. Nørrung, L. Robertson, G. Ru, M. Sanaa, M. Simmons, P. Skandamis, E. Snary, N. Speybroeck, B. Ter Kuile, H. Wahlström, K. Baptiste, B. Catry, P. S. Cocconcelli, R. Davies, C. Ducrot, C. Friis, G. Jungersen, S. More, C. Muñoz Madero, P. Sanders, M. Bos, Z. Kunsagi, J. Torren Edo, R. Brozzi, D. Candiani, B. Guerra, E. Liebana, P. Stella, J. Threlfall, and H. Jukes. 2017. "EMA and EFSA Joint Scientific

Opinion on measures to reduce the need to use antimicrobial agents in animal husbandry in the European Union, and the resulting impacts on food safety (RONAFA)." *EFSA Journal* 15 (1):4666. doi:10.2903/j.efsa.2017.4666.

Na. 2001. "Shattuck lecture: Neurodegenerative diseases and prions." *Infectious Diseases in Clinical Practice* 10 (5):293. doi:10.1097/00019048-200106000-00030.

Nathanson, N., J. Wilesmith, and C. Griot. 1997. "Bovine spongiform encephalopathy (BSE): Causes and consequences of a common source epidemic." *American Journal of Epidemiology* 145 (11):959–969. doi:10.1093/oxfordjournals.aje.a009064.

Nosratimovafagh, A., A. E. Fereidouni, and F. Krujatz. 2022. "Modeling and optimizing the effect of light color, sodium chloride and glucose concentration on biomass production and the quality of arthrospira platensis using response surface methodology (RSM)." *Life* 12 (3). doi:10.3390/life12030371.

Ray, B. 1992. "Cells of lactic acid bacteria as food biopreservatives." In: M. A. Ray Daeschel (Ed.), *Food Biopreservatives of Microbial* Origin (pp. 1–20). CRC Press.

Ray, B., and W. E. Sandine. 1992. Acetic, propionic and lactic acids of starter culture bacteria as biopreservatives. In: B. Ray and M. A. Daeschel (Eds.), *Food Biopreservatives of Microbial* Origin (pp. 103–136). CRC Press.

Robbins, J. 1987. *Diet for a New America.* Stillpoint.

Rogers-Brambell, F. W. 1965. *Report of the Technical Committee to Enquire into the Welfare of Animals Kept Under Intensive Livestock Husbandry Systems.* Her Majestys Stationery Office.

Sapeika, N. 1973. "Food additives." *World Review of Nutrition and Dietetics* 16:334–362. doi:10.1159/000393597.

Serrano, P. H. 2005. *Responsible Use of Antibiotics in Agriculture* (Technical Paper). FAO Fisheries.

Sharma, C., N. Rokana, M. Chandra, B. P. Singh, R. D. Gulhane, J. P. S. Gill, P. Ray, A. K. Puniya, and H. Panwar. 2018. "Antimicrobial resistance: Its surveillance, impact, and alternative management strategies in dairy animals." *Frontiers in Veterinary Science* 4:237.

Sharma, N. K., M. Choct, S. Wu, and R. A. Swick. 2017. "Nutritional effects on odour emissions in broiler production." *World's Poultry Science Journal* 73 (2):257–280. doi:10.1017/S0043933917000046.

Sheweita, S. A., A. M. Batah, A. A. Ghazy, A. Hussein, and A. A. Amara. 2019. "A new strain of Acinetobacter baumannii and characterization of its ghost as a candidate vaccine." *Journal of Infection and Public Health* 12 (6):831–842. doi:10.1016/j.jiph.2019.05.009.

Singh, S., and R. Shalini. 2016. "Effect of hurdle technology in food preservation: A review." *Critical Reviews in Food Science and Nutrition* 56 (4):641–649. doi:10.1080/10408398.2012.761594.

Sridhar, A., M. Ponnuchamy, P. S. Kumar, and A. Kapoor. 2021. "Food preservation techniques and nanotechnology for increased shelf life of fruits, vegetables, beverages and spices: A review." *Environmental Chemistry Letters* 1715–1735. doi:10.1007/s10311-020-01126-2.

Sturino, J. M., M. Rajendran, and E. Altermann. 2013. "Draft genome sequence of the pediocin-encoding biopreservative and biocontrol strain pediococcus acidilactici D3." *Genome Announcements* 1 (3). doi:10.1128/genomeA.00208-13.

Taylor, N., and T. D. Signal. 2009. "Willingness to pay: Australian consumers and "on the farm" welfare." *Journal of Applied Animal Welfare Science* 12 (4):345–359. doi:10.1080/10888700903163658.

Truong, D. B., N. V. Cuong, P. H. Doan, N. T. T. Dung, B. T. Kiet, J. Rushton, and J. Carrique-Mas. 2021. "Small-scale commercial chicken production: A risky business for farmers in the Mekong Delta of Vietnam." *Preventive Veterinary Medicine* 195:105470. doi:10.1016/j.prevetmed.2021.105470.

Wang, B., W. Chen, C. Sa, X. Gao, S. Chang, Y. Wei, J. Li, X. Shi, L. Zhang, C. Zhang, W. Li, and H. Sun. 2024. "Dynamics of antibiotic resistance genes and the association with bacterial community during pig manure composting with chitin and glucosamine addition." *Frontiers in Microbiology* 15:1384577. doi:10.3389/fmicb.2024.1384577.

Xiong, W., Y. Sun, and Z. Zeng. 2018. "Antimicrobial use and antimicrobial resistance in food animals." *Environmental Science and Pollution Research* 25 (19):18377–18384. doi:10.1007/s11356-018-1852-2.

Yigit, A. A., A. K. Panda, and G. Cherian. 2014. "The avian embryo and its antioxidant defence system." *World's Poultry Science Journal* 70 (3):563–574. doi:10.1017/S0043933914000610.

Index

For Product Safety Concerns and Information please contact our EU
representative GPSR@taylorandfrancis.com
Taylor & Francis Verlag GmbH, Kaufingerstraße 24, 80331 München, Germany

www.ingramcontent.com/pod-product-compliance
Ingram Content Group UK Ltd.
Pitfield, Milton Keynes, MK11 3LW, UK
UKHW020205200926
475702UK00004B/616
9781032590745